The Genetic Revolution

The Genetic

Scientific Prospects
and Public Perceptions

A Study Sponsored by the American
Academy of Arts and Sciences

Revolution

Edited by Bernard D. Davis

The Johns Hopkins University Press
Baltimore and London

The Johns Hopkins University Press
701 West 40th Street, Baltimore, Maryland 21211-2190
The Johns Hopkins Press Ltd., London

Library of Congress Cataloguing-in-Publication data has been moved to last
printed page of book.

This collection of essays was prepared under the auspices of the American
Academy of Arts and Sciences with the support of grants from the John D. and
Catherine T. MacArthur Foundation, the Lucille Markey Charitable Trust, and the
Rockefeller Foundation. Any opinions, findings, conclusions or recommendations
expressed herein are those of the individual authors and do not necessarily reflect
the views of the American Academy or the supporting foundations.

Contents

Foreword

The "genetic revolution" has been a continuous one for at least the 125 years since Darwin and Mendel. The scientific arguments for human affinities to, and descent from, the lesser animals sparked a theological debate that still bursts into bitter recrimination in the United States. Indeed, it has been turned on its head by the advocates of equal rights for our animal cousins.

Modern advances have brought genetic science into closer confluence with technology through an ever-expanding capacity to modify organisms—viruses, bacteria, crop plants and animals, perhaps humans. The focus of debate has shifted to more pragmatic concerns over policy and ethical limits. What range of technological productions warrants regulatory oversight? What processes should govern that regulation? How should we balance the benefits and risks? In what forum?

Some thirty years ago I began my own appeals for more public and political attention to the policy implications of genetic engineering.[1] I was naive enough to expect that there could be measured policy analysis in the abstract, some years (but not many decades) before the practi-

---

1. J. Lederberg, "Biological Future of Man," in *Man and His Future: Ciba Found. Symp. 1962*, ed. G. Wolstenholme (Boston: Little, Brown, 1963), pp. 263–73.

cal realization of the emerging techniques. Whatever debate ensued was not gratifying. It was colored more by the imagery of Frankensteinian monsters than by technical assessment of the consequences of alternative choices.

The conference on which this book was based was a fresh start at honest dialectic, seeking synthesis and persuasion. One may at least hope that some perspectives were illuminated, even if not drastically revised, by mutual exposure in this temper. The representation of views was remarkably broad, though understandably lacking in vehemently committed Luddites compared with technological optimists. The protagonists also had much to learn from the sociologically and politically seasoned observations of Aaron Wildavsky and Harvey Brooks, who have chronicled the irrational factors that have driven many a policy debate. In many capitals, it is said that policy is politics.

The convener and editor, Bernard D. Davis, is an unregenerate optimist, in regard to both the likelihood of benefit from new technologies and the role of rational argument in determining social policy. I do not always agree with his conclusions, but I have great admiration for his patience, diligence, and commitment to that process. This by definition entails a drawing out of views, delineation of detail, meticulous discussion, and, if need be, point-by-point rebuttal of views from many sources.

This is not the place for another chapter detailing my own views on the debate. I decided fifteen years ago that I would only be repeating myself were I to continue adding my voice, and the pleasure of offering insights should be shared with others.[2] Just a few brief comments:

Among the choices presented, case-by-case assessment makes the most sense, at least at this stage, to me. We need a lot of pragmatic experience to learn what we mean by our abstract principles, how much we are willing to pay for forgone short-run benefits, who pays and who benefits. Our understanding of the moral aspects is also changing rapidly from one generation to the next, and especially as the world grows ever more crowded. All this creates a lot of trouble, and some innovations may be slowed; worse would be the premature adoption of fixed rules equally likely to be too restrictive for later generations.

2. J. Lederberg, "Biological Innovation and Genetic Intervention," in *Challenging Biological Problems—Directions toward Their Solution*, American Institute of Biological Sciences 25th Anniversary volume, ed. J. A. Behnke (New York: Oxford University Press, 1972), pp. 7–27; J. Lederberg, "DNA Research: Uncertain Peril and Certain Promise" (1975), in *The DNA Story: A Documentary History of Gene Cloning*, ed. J. D. Watson and J. Tooze (San Francisco: Freeman, 1981), pp. 56–60.

We cannot avoid taking some risks. Perhaps we could achieve better-balanced, more consistent policies by noting the tradeoffs we have already elected in related fields. For example, we do not try to minimize the risk of spreading a grave disease by managing all infectious disease patients at a P-4 level of containment, even though we cannot tell what a momentarily undiagnosed fever might portend. Neither do we place all international travelers under quarantine, though such travelers have already spread several pandemics of measles, influenza, and HIV. (There is no point in quarantine after a virus has traversed international boundaries.) But the levels of risk that can reasonably be inferred for considered environmental introductions of engineered organisms are generally much lower than what we accept in the name of ordinary global discourse. It is not that "natural" is necessarily better: microbial nature is surely not benign! But we accustom ourselves to living with nature, and to the scale of risks that this unavoidably entails.

We might use more homely omens, drawn from the everyday behavior of our contemporaries, instead of the social dystopias invoked in some futurologies. Genetic technologies could add very little to the brutalities wreaked by a Hitler or a Saddam Hussein with bullets and fumigants. Much more pressing problems might arise from unbridled technical success in giving people exactly what they yearn for. How do we prepare for another doubling of the life span? And may not the most likely abuse of biotechnology be a parallel with what we have seen in psychopharmacology: self-administered highs, achieved with one-shot neurotropic gene implants?

Joshua Lederberg
University Professor
Rockefeller University

Preface

In a world surfeited with hyperbole the word *revolution* readily evokes skepticism. But if it is ever a useful term, to characterize a fundamental change in our outlook or our powers, it surely applies to molecular genetics as much as to the earlier agricultural, industrial, Copernican, and Darwinian revolutions.

The genetic revolution was foreshadowed by two great discoveries. After decades of classical genetics, which could deal with genes only as purely formal units of inheritance, in 1944 Oswald Avery, Maclyn McCarty, and Colin MacLeod showed, most unexpectedly, that the material substance of the gene is deoxyribonucleic acid (DNA). Nine years later James Watson and Francis Crick launched the field of molecular genetics by discovering that DNA is composed of two complementary strands—a remarkable structure that immediately accounted for the ability of genes to replicate information in successive generations. The rapid growth of molecular genetics then yielded, within a mere twenty years, the revolutionary ability to alter genes at will in the test tube, and to insert them into any organism.

This development, called genetic engineering, has provided enormously powerful tools to virtually every field of biological and medical research and to a rapidly expanding biotechnology industry. In addition, in the study of evolution—the unifying theme of the life sciences—

molecular genetics provides deep new insights into mechanisms, as well as powerful new ways to determine and to time the branching lines of descent in the 4 billion years between bacteria and man.

However, along with great expectations, genetic engineering has generated concern about our ability to handle the new manipulative and predictive powers wisely, and also anxiety about possible risks. Here the public is treading on unfamiliar ground, where it finds it hard to know whom to trust. In particular, the introduction of genetically engineered organisms into the environment, for agricultural or related purposes, has given rise to a most unusual controversy within the scientific community. Microbiologists and molecular biologists are eager to test promising recombinants derived from familiar, harmless organisms; but for many ecologists, concerned with any changes in the distribution of organisms in the environment, the novelty of the recombinants is a cause for concern that would justify considerable delays in testing. Although all agree that the risks have a low probability, the differences in emphasis have made it difficult for our regulatory agencies to agree, even after several years, on a consistent set of regulations, or on definitions for harmless classes of organisms that might be exempted.

In its applications to humans, molecular genetics raises different kinds of problems. One is defining the moral limits of gene therapy and its possible shading into eugenics. Another is created by our increasing ability to detect genetic susceptibility to various diseases. For this knowledge, intended to benefit individuals, may also be used to invade their privacy, or to burden them with discouraging predictions that might be unwelcome.

This book is intended to provide assessments by highly qualified experts on various aspects of the problems raised by the genetic revolution. Since the problems have already been extensively debated, it seemed that any further contribution, to be valuable, would require highly professional, sophisticated analyses of the various topics. To lay readers the prospect of such a book, with chapters by many different authors, may seem discouragingly academic. We have therefore made extra effort to ensure a style that is accessible.

The book is organized in the following way. The first chapter discusses the key issues. The next, on the origins of molecular genetics, portrays a remarkable intellectual triumph, of interest in its own right and as a model for future growth of the field. In addition, that chapter provides a vocabulary for the subsequent scientific chapters, which discuss where we now stand, and where we are likely to go in the near future, in applying the new techniques—to microbes, plants, animals, humans, and the understanding of evolution.

The rest of the book focuses on public perceptions and governmental responses, and on the factors that influence them. These chapters provide a variety of perspectives, coming from environmentalism, the law, political science, and governmental regulation. But along with some conflicting views there is agreement that wise policy must balance the likelihood and the magnitude of the benefits and the dangers, rather than seek an unattainable zero risk. There is also agreement on the need for better public education on the relevant science—especially recognition that only a tiny fraction of microbial species cause disease, while the vast majority have beneficial and even indispensable roles in the cycle of life.

Two closing chapters provide a summary, with critical commentary. For those readers who find some of the scientific chapters too detailed, the summary of their high points in the chapter by the editor (Chapter 14) will provide an adequate replacement as a background for the chapters on policy. In the final chapter Harvey Brooks, a physicist long engaged in problems of public policy, comments on the chapters that bear particularly on this aspect of the subject. This discussion will bring out interesting similarities between policy problems now encountered in genetic engineering and those that arose earlier from the physical sciences.

Finally, although the controversies over the regulation of engineered organisms are still not settled, there is little doubt that biotechnology will flourish. Yet it is also likely to face continuing wasted effort and delay, as it tries to accommodate to public perceptions that do not distinguish real problems from excessively remote possibilities. We hope this book will illuminate these issues for a wide audience of concerned citizens, as well as biomedical scientists and government officials. It may also provide a useful paradigm for future novel intersections between science, technology, and society.

The American Academy of Arts and Sciences sponsored this book, including a conference of the contributors. The book is in a sense a specialized continuation of an earlier publication of the Academy, *Progress and Its Discontents* (edited by G. A. Almond, M. Chodorow, and R. H. Pearce, University of California Press, 1982). Related topics have also been taken up in several issues of the Academy quarterly *Daedalus*: Science and Its Public: The Changing Relationship (Summer 1974); Limits of Scientific Inquiry (Spring 1978); Modern Technology (Winter 1980); Scientific Literacy (Spring 1983); America's Doctors (Spring 1986); Art and Science (Summer 1986); and Risk (Fall 1990).

We are very grateful for generous financial support from the Lucille

P. Markey Charitable Trust, the MacArthur Foundation, and the Rocke-feller Foundation. The editor is indebted to Jacqueline Wehmueller and Carol Zimmerman at the Johns Hopkins University Press, and to Alex-andra Oleson at the American Academy of Arts and Sciences, for skill-ful and pleasant cooperation. Virginia LaPlante and Mary Yates pro-vided invaluable editorial advice.

Contributors

*Charles J. Arntzen, Ph.D., Deputy Chancellor for Agriculture, Texas A and M University

*Francisco J. Ayala, Ph.D., Donald Bren Professor of Biological Sciences, University of California at Irvine

*Harvey Brooks, Ph.D., Professor, John F. Kennedy School of Government, Harvard University

*Allan M. Campbell, Ph.D., Professor of Biology, Stanford University

*Bernard D. Davis, M.D., Adele Lehman Professor of Bacterial Physiology Emeritus, Harvard Medical School

Theodore Friedmann, M.D., Professor of Pediatrics and Genetics, University of California at San Diego School of Medicine

Rollin B. Johnson, Ph.D., Project Director, John F. Kennedy School of Government, Harvard University

*Simon A. Levin, Ph.D., Charles A. Alexander Professor of Biological Sciences, Cornell University

Franklin M. Loew, D.V.M., Ph.D., Dean and Foster Professor of Comparative Medicine, Tufts University School of Veterinary Medicine

Margaret Mellon, Ph.D., National Wildlife Federation

Henry I. Miller, M.D., Director, Office of Biotechnology, U.S. Food and Drug Administration

*Member, National Academy of Sciences

James H. Schwartz, M.D., Ph.D., Professor of Physiology and Cellular Biophysics and of Neurology, Center for Neurobiology and Behavior, Howard Hughes Medical Institute, Columbia University College of Physicians and Surgeons

Richard B. Stewart, Assistant Attorney General for Environment and Natural Resources, U.S. Department of Justice, and Professor, Harvard Law School (on leave)

Aaron Wildavsky, Ph.D., Professor of Political Science, University of California at Berkeley

1

The Issues: Prospects versus Perceptions

Bernard D. Davis

The seed of the genetic revolution was planted when Oswald Avery and his colleagues identified the genetic material as deoxyribonucleic acid (DNA) in 1944. The plant began to emerge as the growing field of molecular genetics in 1953, when James Watson and Francis Crick elucidated the fundamental structure of DNA. The revolution then flowered with the development, in the 1970s, of a technique that permitted unlimited manipulation of DNA: *molecular recombination*.

In fact, recombination between DNA chains occurs in nature within cells as a regular feature of sexual reproduction, allowing every germ cell (sperm or egg) to inherit a different, randomly selected set of genes from the double set of the parent. What the term *recombinant DNA* now refers to is the product of recombination in the test tube between two DNA chains, not at random but at specific sites within each chain.

The development of recombinant DNA offers profound lessons on the nature of scientific advance. Rather than being a single great discovery, it emerged from the synthesis of a number of earlier discoveries, each initially quite esoteric. Moreover, it was quite unpredictable, like most highly original discoveries that open up new fields. Finally, the discovery of molecular recombination illustrates the long delay that often occurs before basic research yields major payoffs—a major point, since legislators had understandably been impatient that so few practical applications had emerged after two decades of increasing support of molecular genetics.

Along with procedures that made it possible to splice together fragments of DNA from any sources, techniques were developed for incorporating the product into a host bacterium, thus making it possible to obtain a limitless number of identical copies. Geneticists use the term *clone* for a set of genetically identical individuals, such as plant cuttings or asexually multiplying bacteria; hence, the multiplication of an inserted gene as part of the multiplying cells in a bacterial culture is called *cloning* of that gene.

This development has opened up remarkable possibilities. Among the practical ones, recombinant cells can be used to manufacture valuable products, such as human insulin formed inexpensively in bacteria. And similar incorporation of added genetic material (called a *transgene*) into a higher plant or animal may make the organism itself more valuable, such as a tomato plant made resistant to a herbicide or more flavorful. But even more important has been the impact on fundamental research, based not only on cloning genes but also on modifying them at will.

With these and related tools investigators are now successfully attacking the biochemistry of processes that formerly seemed far too complex. One example is memory (which is discussed in Chapter 10 of this volume). Another is differentiation, the process by which a fertilized egg cell gives rise to the many different kinds of cells and organs of the developed organism. We cannot foresee how far our insights into the human organism will go, or the extent of the benefits—and the problems—that these powers will also give us.

Students of the history of ideas might say that all these advances, though impressive, still do not meet the criterion of a scientific revolution, as defined by philosopher of science Thomas Kuhn.[1] He emphasized that "normal science" progresses through cumulative steps within a generally accepted conceptual scheme, called a *paradigm*; but however large some of these steps, he would use the term *revolution* only for discoveries that overthrow a paradigm. But while this widely accepted view fits twentieth-century physics, where relativity and quantum theory had a radical impact on commonsense paradigms about the nature of time, space, and matter, the great advances in biology have generally filled a vacuum rather than overthrown a paradigm. Molecular genetics, however, meets even Kuhn's criterion, for it has challenged two paradigms: the assumption that mutations can occur only randomly and cannot be directed, and the assumption that our knowledge about the tempo and the pattern of evolution must come from indirect evidence (the comparison of visible traits among fossils and among living species). DNA sequences, the very stuff of evolution, now provide direct measurements of evolutionary relationships between different organisms and between different genes within an organism (Chapter 11).

As a curious linguistic aside we may note that the terms *genetic* and *engineering* have the same etymological origin. The word *genetic* comes directly from the Greek and Latin roots for birth. *Engineering* has a more tortuous derivation. *Genius* originally meant an inherited nature, as in the genius of a people or a language; it came to mean a nature with unusual intelligence; the noun gave rise to the adjective *ingenious*; and an ingenious device or machine became an *engine*—designed or run by an engineer.

To evaluate the problems that this book will discuss, the reader obviously must have some basic knowledge of molecular genetics. Yet we have not considered it necessary to provide a detailed primer. The reason is that much of the extraordinary complexity of living beings can be broken down into the intricately organized interactions of their component parts, analyzed at successive levels of refinement. Indeed, the term *organism*, introduced in the nineteenth century, reflects this overriding principle of organization in living beings, then observed at the level of organ structure and physiology; and it applies equally at the levels of biochemistry and genetics. Fortunately, even more than some other aspects of organisms, the basic principles of molecular genetics can be adequately presented in remarkably simple and mechanical terms, without requiring the reader to master a large amount of detail or even to visualize specific chemical structures. We shall define most of the key genetic and biochemical terms in Chapter 2, and others as they appear in later chapters.

Public Concerns

The public has been concerned about several conceivable risks from the growth of molecular genetics, some likely and others very remote. Among these, the applications to humans can raise serious moral problems, especially if the procedures should be extended from the medical goal of preventing or treating disease to the nonmedical, eugenic purpose of enhancing desired traits. Another important problem is the possibility of social or psychological harm in the use of genetic information with predictive implications for the future health of an individual. In dealing with these problems a public consensus must play a large role, and the public needs a realistic picture of what is likely or possible and what is remote or even impossible.

With genetically engineered bacteria or other organisms, however, the problem is primarily a technical one: detecting or estimating the likelihood of physical harm. Here the public cannot usefully play a similar role. The many years of experience with recombinant DNA, spreading now even to high school laboratories, have so far not led to any detectable harm, and many scientists are convinced that there will be none (unless one is dealing with DNA from effective agents of disease). One must therefore ask why the public, and government regulators, continue to be so concerned.

One reason, to be spelled out in Chapter 3, is that the molecular geneticists who first developed recombinant bacteria containing foreign genes were themselves relatively unfamiliar with the microbial world. They not only underestimated the difficulty in converting a harmless organism into an effective pathogen. They also exaggerated the novelty of the transfer of DNA between bacterial species; such transfers are now increasingly seen to occur in nature. Although most scientists in the field quickly got over their initial anxiety, the public understandably assumed that if the scientists were conceding so much danger the reality must be even worse.

A broader base for public anxiety undoubtedly stems, by extrapolation, from our disillusioning experience in recent decades with the physical technologies. After a long period of expecting a free lunch, we have recognized belatedly that the benefits of these technologies have often been accompanied by large costs—for example, pollution of the environment, exhaustion of natural resources, and even damage to the atmosphere. It has been tempting to assume that similar unintended and unforeseen harmful consequences are also inevitable for biotechnology. And indeed, no one can provide an absolute guarantee against such eventual problems.

Nevertheless, the base for this extrapolation from the physical to the biological technologies is weak. For biotechnology also has a long history; but its balance of benefits and harms is quite different from that of the physical technologies. The name given to the earlier biotechnology was *domestication:* the adaptation of initially wild organisms, by empirical genetic selection, to meet human needs. Domestication has provided enormous benefits. These include animal and vegetable foods of greatly increased abundance, variety, and quality; fibers and leather; the beauties of horticulture; the gratifying and even ennobling interaction with pets; and beverages and bread fermented by selected yeasts. Genetic engineering will increase the speed, precision, and range of domestication. However, the purpose will remain the same: to strengthen those traits, in animals, plants, or microbes, that make the organism more useful to us.

The important point here is that the benefits of domestication have been remarkably free of harmful side effects. There is no obvious reason to expect that expansion by the techniques of genetic engineering will have different consequences. In particular, in evaluating the likelihood of inadvertent, uncontrollable spread of engineered organisms we should note that domestication usually makes organisms less well adapted to the wild, and even totally dependent on human beings for their propagation. More specifically, while some domesticated animals can indeed revert to wild, feral life (such as wild horses, or cats in cities), there are no known cases in which they are better adapted than their wild ancestors to the same natural environment.

Accordingly, since genetic engineering of animals and plants is used to introduce even larger or more foreign hereditary changes than those appearing in nature by mutation or recombination, the products are likely to remain restricted to the geographical areas where we cultivate them, rather than to spread beyond control. The real problem, then, would not be their escape, but would be the same as that of existing agriculture: spread of domesticated crops and animals not spontaneously by our deliberate cultivation. The benefits are therefore accompanied by such costs as increased displacement of natural wild areas, erosion of the soil, and exacerbation of the population problem.

Fear of Microbes

The greatest present controversy is over the intentional introduction of engineered bacteria into the environment. (This procedure is often called *deliberate release* in this country, though not in England; but since that term seems to imply that the material should have been held back, most contributors to this book have accepted the more neutral term *planned introduction*.) Because the presumed dangers here are invisible and intangible, it is particularly difficult for the public to assess the contradictory judgments that it encounters. Common sense and past experience are not very useful guides.

In addition, the public starts with a prejudice based on its familiarity with bacteria mostly as dangerous "germs," causing disease. In fact, such pathogens are only a tiny, specialized fraction of all bacteria. The vast majority of microbes (bacteria, molds, and protozoa), in the soil and in water, play a role that is not only valuable but essential, in recycling dead organic matter into a form that can be taken up again in living organisms. We need a good deal of public education on the real nature of the microbial world.

Although there is thus exceedingly little scientific basis for expecting recombinants of harmless bacteria to present dangers, negative perceptions promoted by prophets of doom have received much attention in the press. And since government officials must take into account public perceptions as well as technical assessments, their official responses have tended to be cautious. Meanwhile, virtually all microbiologists and molecular biologists now agree that regulations should be based on the properties of the organism—whether potentially dangerous or likely to be harmless—rather than treating variants produced by genetic engineering as a special class because of the technique used to obtain them. But as we shall see in Chapter 12, regulatory agencies still have great difficulty in turning this principle into practice—especially the office most focused on the environment, the Environmental Protection Agency.

The Perspectives of the Authors

The authors did not seek consensus on controversial issues. Some would accept the public perceptions as a given, which regulators must consider as important as the scientific evidence. Most would build policy primarily on technical assessment by qualified scientists, even in an era when many people mistrust experts and institutions, and they would emphasize that the government should play a more active role in educating the public about misconceptions on the scientific issues.

Nevertheless, the problem of scientifically assessing risk is not always straightforward. An empirical approach, acquiring sufficient data about the source of potential hazard, has given satisfactory answers when applied to definite, quantifiable risks, such as chemicals with a rapid toxic effect. But it has not done well with negative predictions, such as absence of a hypothetical risk from a novel organism or material. Here risk assessment must rely, if only tacitly, not only on empirical data but also on extrapolation from general scientific principles.

For judging environmental risks from recombinant bacteria, these principles come mainly from two sources: our knowledge of the microbial world, and our understanding of evolutionary biology—the discipline that deals with factors influencing the survival and spread of competing organisms in nature.[2] Evolutionary biology therefore figures more prominently in this book than in most discussions of the problems of genetic engineering.

In addition to considering how to assess specific risks, some chapters consider the broader problem of how to formulate policy. The authors agreed, in principle, that a realistic policy must deal with probabilities, rather than seeking an unattainable absolute security. There was less agreement, however, on how to treat distant or very hypothetical risks that have aroused public concern.

In the end the authors tended not to look far into the future, perhaps because the history of forecasts of science and technology has invariably revealed a clouded crystal ball. The scientific chapters in this book therefore focus mostly on near-term opportunities, powers, and dangers. Indeed, as the most extensive analysis of policymaking in the book (Chapter 6) suggests, when dealing with new technologies our assumption that we should try to anticipate problems is rarely very productive, even in the short term. A more realistic and effective approach would instead seek "resilience": that is, trying to recognize newly arising problems rapidly as they arise, and to respond vigorously. Moreover, however much we may wish to act wisely, we surely cannot foresee where the genetic revolution will take us in the decades and the centuries that lie ahead, nor can we solve these future problems for posterity.

While the contributors to this book disagreed somewhat about the assessment of risks, they did agree on one feature of the public debate: the need to combat the influence of those critics, such as Jeremy Rifkin, who are ideologically opposed to all genetic engineering (though in specific cases they often present their objections in more moderate terms). Their appeal to public uneasiness about technology in general, and about the unfamiliar microbial world in particular, has made them highly newsworthy; and they have had a good deal of political influence, especially the Green political party in Europe. Unfortunately, scientists are generally unskilled at public relations and find it difficult to deal with professional activists. The public is therefore often exposed to a distorted picture that is only slowly corrected.

In a thoughtful essay, Freeman Dyson has emphasized that saying no to a new technology is generally safer for regulators than saying yes, but it often has larger costs for society.[3] Indeed, in the long run we really have no choice but to try to channel and control new technologies, rather than to avoid problems by blocking them. Our dense world population is now extraordinarily dependent on technology for its survival, and blocking further advances, out of guilt over the uneven distribution of the costs and benefits, will hinder rather than help to solve this serious social problem. Even more, we cannot unlearn the scientific method—which could only have been stopped before Galileo—nor can we repress the parts of our evolutionary heritage that underlie it: curiosity, and the urge to take advantage of the most efficient means for meeting our material needs or desires.

A serious approach to policy must also recognize the limits of science, as well as its positive contributions. The strength of science lies in its systematic search for objective knowledge about a restricted set of problems: those aspects of the natural world for which there are objective answers. But science cannot prescribe social policies, because they involve value judgments and hence cannot be altogether objective.

Nevertheless, science can be a reliable critic of the underlying assumptions to which the values are applied, and so it can help us to avoid building policy on false or shaky assumptions. In the gray areas where the science cannot be definitive—as when it tries to predict the properties of a class of yet unknown organisms—we may differ, as individual scientists and as citizens, in our interpretations of the evidence, and in our judgment about how much to rely on established principles, just as we differ, often with great intensity, over values, goals, and political preferences. But we must try scrupulously not to let these differences about such subjective matters distort our interpretations and judgments, as we try to recognize the objective knowledge that science does eventually provide.

Public suspicion of genetic engineering not only has had a direct effect on the advance of biotechnology; it has also added one more facet to a more general skepticism about the goals, and the social impact, of science and technology. As the sociologist Edward Shils warned a few years ago, this antiscience movement poses a threat, more than is generally recognized, to public support of science, the recruitment of promising students, and ultimately the morale of working scientists.[4] Events have confirmed each of these dire predictions, perhaps the best documented being the decline of student interest in science and the abysmal performance of American students in international tests. In the next decades we shall surely pay a price, in both our culture and our economic health. The portrayal of engineered organisms as potential monsters fosters the widespread negative image of science and scientists. We hope this book will help to provide a more realistic perspective.

Notes

1. T. S. Kuhn, *The Structure of Scientific Revolutions*, 2d ed., enlarged (Chicago: University of Chicago Press, 1970).

2. B. D. Davis, "Evolutionary Principles and the Regulation of Engineered Bacteria," *Genome* 31 (1989): 864.

3. F. J. Dyson, "On the Hidden Costs of Saying 'No,'" *Bull. Atomic Sci.* 31 (June 1975): 23.

4. E. Shils, "The Antiscience Movement," in B. D. Davis, *Storm over Biology* (Buffalo: Prometheus Books, 1986), Introduction.

2

The Background: From Classical to Molecular Genetics

Bernard D. Davis

Though molecular genetics was launched by the discovery of the structure of DNA in 1953, its explosive advance was made possible in large measure by another development, which occurred during the preceding decade: the emergence of bacterial genetics. This advance drew bacteria and their viruses—the smallest and simplest forms of life—into the fold of genetics, and these organisms then proved to be ideal for studying genes at a molecular level. In addition, because bacteria were best known as agents of disease, and because they were the first organisms to be genetically engineered, they later became the center of the largest controversy over the applications of these new techniques. For these reasons this chapter will review key features in the development of bacterial genetics as well as in classical and molecular genetics.

The chapter has several purposes. One is simply to tell the story of an outstanding intellectual achievement. Another is to serve as a sort of glossary, using the history of the field to introduce technical concepts and terms that will appear repeatedly in later chapters. Still another is to consider the influence of molecular genetics on some philosophical concepts. And finally, the path to recombinant DNA will illustrate significant features of the process of research—in particular, the fact that major discoveries often arise in unexpected ways and cannot be forced.

The reader who is already familiar with genetics may wish to skip

the earlier part of this chapter, but he or she may find some novel interpretations in the last section, on conceptual contributions and philosophical challenges of molecular genetics.

Classical Genetics

The basic principles of genetics were recognized in 1863 by Gregor
Mendel, an Austrian monk who studied inheritance in peas in the
garden of a monastery in Brünn. Each trait that he studied was found to
be transmitted from parents to progeny as though it were determined by
a discrete, independently inherited unit, later called a *gene*. But his
mathematical approach, comparing the frequency of a trait in parents
and offspring, was then unfamiliar in biology, and so his discovery was
ignored. Not until it was rediscovered, in 1900, did it initiate genetics as
a science—that is, as a growing, cumulative body of knowledge about
the nature of the external world.

Geneticists inferred the existence of genes from differences in a
trait among related individuals, reflecting the presence of different, alternative forms of the gene, called *alleles*. Initially they dealt, like Mendel, only with the variations that nature provided; but later investigators broadened the range by experimentally isolating *mutations:* rare,
random changes in a gene, whose frequency can be increased by radiation or by mutagenic chemicals. Molecular studies of the mechanism
showed that mutations arise through many different kinds of changes in
DNA sequence, including replacement of a *nucleotide* (one of the four
kinds of units in the chain of DNA), deletion or insertion of a nucleotide or of a longer set of nucleotides, or rearrangement or transposition
of a block of DNA within an organism's *genome* (the total of its genes).

Genes are located at specific sites on *chromosomes*—bodies of
DNA associated with protein in the cell nucleus, and visible under the
microscope at certain stages of cell division as threadlike structures.
Their movements in cell division are guided to ensure the equal distribution of the duplicated set of genes to the two daughter cells. A major triumph of classical genetics was learning how to map the sequential
location of the genes on a chromosome, in the following indirect way.

In all the cells of an animal or plant the chromosomes are found in
pairs, with one member from each parent. The two members are *homologous*—that is, they contain the same sequence of genes, concerned
with the same set of traits, but often with slight, allelic differences. In
the formation of *germ cells* (egg or sperm cells, also called *gametes*) each
pair of homologous chromosomes undergoes random *recombination;*
that is, a new chromosome is made partly from one member of a pair

and partly from the other, thus giving rise to a unique new combination of genes.

In this process the two parental chromosomes lie side by side, and as the molecular copying machinery moves along them it *crosses over* from one to the other, with a roughly equal probability at any location. Accordingly, the frequency of inheriting two genes together from the same parent (*linkage*) provides a measure of the distance between them: the shorter the distance, the greater the linkage. Determining the order of identifiable genes (or other identifiable sites) on a chromosome yields a *linkage map*.

If two genes are sufficiently far apart on a chromosome, or if they are on separate chromosomes, they are not linked at all but are independently inherited—that is, each has an equal chance of coming from either parent. Mendel was fortunate in studying half a dozen genes that all happened to be unlinked; otherwise the transmissions of traits from parents to progeny that he observed would not have had simple ratios, and the data would have been too confusing to lead to his fundamental laws. This kind of experience is not rare in science: an important principle is discovered through the simplest cases and must then be substantially modified as exceptions appear. During the transition the exceptions often undermine confidence in the whole model; but the resulting debates are usually settled by further experiments, rather than by the argumentation that prevails in most other areas of intellectual activity.

Mendel's most profound insight was that different *genotypes* (the pattern of genes) could yield the same *phenotype* (the observed set of traits). The reason is that one allele of a gene was *dominant* over another, called *recessive*. That is, *heterozygotes*, carrying both the dominant and the recessive allele, expressed the same trait as a *homozygote* for the dominant allele, carrying it in both chromosomes, while homozygotes for the recessive allele expressed that trait.

Although the concept of dominance was prominent in classical genetics, it has become less important with the development of molecular genetics, since the phenotype can now be identified not only in terms of a visible trait but more precisely in terms of a concrete gene product. Thus, at the molecular level the gene for sickle cell hemoglobin is neither dominant nor recessive: in heterozygotes both the sickle cell and the normal hemoglobin are present in the red blood cells. However, at the level of the disease the gene is recessive, because in heterozygotes the red blood cells function normally, while in homozygotes the cells have only the abnormal hemoglobin and tend to agglutinate in the blood vessels and cause many different symptoms. Heterozygotes for the gene for a human disease are often called *carriers* of the trait.

While genetics started with the simple case of two alleles, a single gene can have a number of different alleles. Indeed, molecular studies have now shown that minor variations in the DNA sequence of a gene are widespread, and while some have large effects on the phenotype, or small effects, most have none ("neutral" mutations). For example, various hereditary anemias are due to different abnormal hemoglobins, each with a very small difference in structure (replacement of 1 amino acid out of 146), but with a disastrous effect on the whole organism. At the other extreme, "silent" nucleotide replacements, which do not affect the function of the protein product, are particularly frequent, since there is no pressure for evolution to select against them (as it does against the vast majority of mutations with phenotypic effects).

Genetics has focused mostly on *monogenic* traits, each governed by a single gene. However, the socially most interesting human traits, such as variations in physique or in behavioral characteristics, are much harder to study, because they are *polygenic* and hence vary quantitatively rather than qualitatively; that is, the trait depends on many different genes, each of which contributes to its total value or its pattern. The fact that genes are important for intelligence or for height does not mean that there is a gene for each of these traits; there are many genes involved, still undefined. Moreover, in most polygenic, quantitative traits the phenotype depends not only on genes but also on environmental factors. The difficulty of analyzing the complex polygenic traits severely limits the possibilities for applying genetic engineering to the socially most interesting human traits in the foreseeable future.

These insights into the nature of heredity have led to many practical benefits. For example, the use of mutagens to increase the rate of random mutation over the spontaneous background rate helped in the isolation of useful new strains, such as antibiotic-producing microbes with higher yields. Genetics had its greatest practical impact by contributing hybrid corn to the "green revolution" (Chapter 7). In this development prolonged inbreeding, which successively decreases heterozygosity in the genome, finally produces strains homozygous for most genes; and while these strains themselves are not highly productive, crosses between certain pairs yield seed for a uniform and very productive crop (hybrid vigor).

Inbreeding has also been helpful to medical research, especially through the development of strains of mice that are homozygous for most of their genes. With these strains, as with identical twins, tissue transplants are not rejected, as they are in the genetically heterogeneous usual strains, and so one can introduce specific genes to identify those that are responsible for rejection. Similarly, inbred strains have greatly aided study of the biology of cancer by allowing cancer cells to be transplanted without causing immunological rejection.

The greatest advances in classical genetics, however, were conceptual. Millennia of speculation about the intriguing problem of biological inheritance had given rise to innumerable myths. For example, earlier literature abounds in anecdotes about prenatal maternal experiences imprinting on the offspring, such as a frustrated desire for strawberries during pregnancy resulting in "strawberry marks." Genetics finally provided a rational framework for understanding inheritance. In one area, however—the relative contributions of heredity and environment to intelligence and other behavioral traits—the imprecision of the answers still generates controversy. It will no doubt persist until the relevant genetic differences can be measured by more direct neurobiological methods, such as comparisons of brain circuitry or of proteins in the synapses (switches) that connect the nerve cells.

Another major conceptual contribution of classical genetics was to close a huge gap in the field of evolutionary biology. Charles Darwin's *The Origin of Species*, in 1859, proposed that evolution occurs by natural selection: the individuals in a species differ in their heritable fitness for a particular environment, and the more fit have more surviving progeny. These graded steps, accumulating in successive generations, were postulated to lead eventually to the formation of new species. But while Darwin presented a mass of evidence for this theory of natural selection, he barely discussed an equally important part of his theory: the requirement for a continuous supply of heritable novelty for natural selection to act on. The reason for his neglect is that the mechanism of this postulated variation was entirely unknown, for there was not yet any science of genetics.[1]

In fact, Darwin entertained as one possibility an idea with much appeal to common sense: that animals can inherit characteristics acquired by a parent in an adaptive response to the environment. Among biologists this idea, often called Lamarckism, lost credibility as genetics advanced. Moreover, molecular genetics has now driven a large nail in the coffin by identifying the mechanism of information flow from DNA sequence to protein sequence (i.e., from genotype to phenotype), for fundamental chemical principles would make it impossible to reverse that flow. Lamarckism is still occasionally revived, however, by people who use philosophical arguments to try to answer this empirical question.

Darwin's theory still evokes much skepticism. The main source is religious objections to its implications for the origin of man. Moreover, people have a strong anthropomorphic tendency to seek deterministic causes for all events, and so they are reluctant to recognize that trial and error—which is the essence of natural selection—can be important in nature. In addition, biologists remained generally skeptical for nearly a century after Darwin, because there was no explanation for the postulated hereditary variation. Not until the 1930s was this missing element filled in, when a marriage between evolutionary biology and genetics generated the answer. Rare mutations, superposed on an ordinarily accurate copying mechanism, produce the needed hereditary novelty—a true genetic innovation, beyond the more limited novelty of the rearrangements resulting from genetic recombination.

Greatly strengthened by this development, evolution became the central, unifying theme of the life sciences. Comparative studies in biochemistry added further support by showing that innumerable molecules and the sequences of enzymatic reactions to make or use them are shared by all organisms, ranging from the primitive bacteria to the latest branches in the tree of evolutionary descent. Conservation of useful molecular devices during evolution now explains this unity.

But more than either genetics or comparative biochemistry, molecular genetics has now offered even more powerful evidence for evolution. As a simple example, if there were no evolutionary continuity between all organisms, it would be hard to explain how research on bacteria can help us to understand brain cells, or how bacteria can translate human genes into useful products. And comparison of DNA sequences goes much farther, offering extraordinarily direct information on the evolutionary connections among organisms.

Evolutionary biology not only has gained from the genetic revolution, but, reciprocally, it also helps us to deal with some of the concerns that have arisen. First, evolutionary principles set limits to how far we can change organisms without wrecking them. In adapting to its environment each species has evolved with a set of genes that are harmoniously fitted to each other, like the parts in a clock. Hence, while we can manipulate DNA at will, our capacity to remake organisms has severe limits. In addition, the evolutionary principles that have governed the spread of competing organisms in nature in the past apply just as much to the organisms now being engineered. For these reasons, genetic engineering gives new importance to public education about evolution, much as nuclear engineering has made the public aware of Einstein's esoteric discovery of the equivalence of mass and energy.

The Delayed Recognition of Genes in Bacteria

Much like classical genetics, the study of microbes (single-celled organisms) was a relatively isolated field for the first four decades of this century. The main branches of the life sciences concentrated on the internal structure and function of organisms, and on their evolutionary connections; and microbes were too small for such studies. They were therefore of interest primarily for their practical impact on their environment: infectious diseases, useful fermentations, and contributions to soil fertility.

Indeed, Darwin showed no interest in the microscopic world being studied by his contemporary Louis Pasteur. Similarly, Pasteur did not recognize that he was dealing with a rapid form of natural selection when he showed that different fermentations are caused by different microbes, which are ubiquitous but are selected by the medium: yeasts that convert sugar to alcohol in grape juice, or bacteria that convert sugar to lactic acid in souring milk. It would have been difficult for anyone to dream then that the monumental advances initiated by Darwin and Pasteur would someday intersect intimately.

Even after classical genetics had begun to flourish, it still had no impact on the study of bacterial variation, for special reasons. For one thing, the usual staining methods did not reveal any chromosome in the tiny bacteria. In addition, these organisms lack sexual reproduction, and so the standard way to demonstrate genes—studying the transmission of alleles between individuals—could not be used. Finally, it was difficult to distinguish physiological adaptation from rapid genotypic adaptation. In the former an environmental stimulus activates a pre-existing useful gene in all the cells of a culture, while in the latter a rare cell undergoes a useful mutation that allows it to grow faster in that environment. The confusion arose from the extremely rapid multiplication of bacteria. Since one cell can become a million overnight, a strongly advantageous mutant can outgrow the rest of the culture, in a form of natural selection, very quickly—far outside the familiar time scale of evolution, and mimicking physiological adaptation. When these problems were overcome bacterial genetics grew rapidly.

At about the same time gene action began to be defined in biochemical terms, through work by George Beadle and Edward Tatum with the mold Neurospora—a microbe with a sexual cycle, and hence available for genetic study. This organism can grow on an extremely simple culture medium, because it makes for itself virtually all of its *metabolites:* compounds, such as amino acids or vitamins, essential for the complex set of metabolic reactions that provide energy for various functions and building blocks for growth. Humans and other mammals, in contrast, have degenerated in their biosynthetic activities, in the course of their evolution, for their diet must provide all of the vitamins and half of the twenty amino acids.

The Neurospora investigators introduced the use of biochemical mutants, each of which had lost the ability to make a particular metabolite and hence required it in the growth medium. A metabolite is synthesized by a sequence of specific enzymes, each of which catalyzes one small chemical change; and in each mutant one reaction in that biosynthetic sequence was blocked. Moreover, there was a one-to-one relationship between the blocked enzymes and the mutated genes. This work provided the first clear indication of the biochemical function of genes: each directs the formation of a corresponding enzyme.

This hypothesis came under much fire, for as is frequent with the initial formulation of broad generalizations in science, exceptions appeared: for example, some genes affected the activity of multiple enzymes. Further discoveries, however, explained the exceptions in terms that required modification rather than destruction of the basic principle. Thus, a regulatory gene, which produces a regulatory protein rather than an enzyme, can affect the activity of many other genes.

Similar biochemical mutants were soon isolated in the common intestinal bacterium *Escherichia coli*. They provided early evidence for the existence of discrete genes in bacteria, and they became major tools in the development of both bacterial and molecular genetics.

The Flourishing of Bacterial Genetics

Meanwhile, the next major advance in bacterial genetics was an unexpected by-product of a lifetime of work by Oswald Avery on the pneumococcus, the main cause of pneumonia in humans. In 1944 Avery, Colin MacLeod, and Maclyn McCarty identified the material of the gene by showing that the transfer of a heritable property from one type of pneumococcal cell to another was mediated by DNA. The designation of this process, *transformation*, is now applied to all transfers of naked DNA from the medium into cells.

Although this discovery planted the seed of molecular genetics, it met with a good deal of skepticism. One reason was that DNA seemed too simple for the complex function of a gene, since it has only four different building blocks, whereas proteins, which are also present in the chromosome, have twenty building blocks. Another reason was the response of a group of investigators who meanwhile had been seeking the nature of the gene, very logically, by working on the simplest possible organisms, *bacterial viruses*, also called *bacteriophages*. But only later were these shown to contain small blocks of DNA (or of RNA, ribonucleic acid, in some viruses), coding for the few genes needed to specify the components of the virus, and using the host cell machinery to make those components. The nucleic acid in the virus particle is surrounded by a specific protein coat that helps the particle to infect a new bacterial host cell. There the DNA multiplies, directs formation of coat proteins for the progeny virus particles, and then releases the complete, infectious particles.

Bacteriophages, being even simpler than bacteria, subsequently played a major role in the development of molecular genetics. Nevertheless, it was hard for the early bacteriophage investigators to admit that they had been scooped, on the fundamental problem of the nature of the gene, by an adventitious discovery made in the course of an essentially medical investigation.[2]

While the discovery of transformation by DNA is justly most famous for initiating molecular genetics, it also helped to launch bacterial genetics, by showing that a gene can be transferred between bacterial cells. Stimulated by this discovery, Joshua Lederberg shifted from medical school to a Ph.D. program and soon discovered two additional mechanisms of gene transfer between bacteria, which proved to be more convenient than transformation. One mechanism is *conjugation*, in which contact between two cells results in formation of a connection through which DNA can move. The other mechanism, discovered with Norton Zinder, is *transduction*, by a bacterial virus as a *vector:* that is, in the course of its assembly in the host cell it picks up a bit of host DNA and transfers that DNA to the cell that it infects. In contrast to the full set of genes transferred in the germ cells in sexual reproduction, the primitive mechanisms in bacteria transfer only a small block of DNA.

Once the fundamentals of bacterial genetics were understood, the very features of bacteria that had hindered physiological and genetic studies became advantages. Huge numbers of individual bacteria can be grown rapidly and in a small volume. And because bacteria are haploid rather than diploid (i.e., they have only a single kind of chromosome instead of homologous pairs), they express alternative forms of a gene without the complication of dominance. But the greatest advantage is that certain rare variants, produced either by mutation or by recombination, can be selected with great efficiency from huge populations. For example, a single streptomycin-resistant mutant, among a billion cells of the parental type, can survive and multiply in a medium containing the antibiotic while all the other cells are killed.

During the nine-year interval between the discovery of the genetic role of DNA and the discovery of its double-helical structure, bacterial genetics matured very rapidly. It was then able to play a large role in the development of molecular genetics.

Foundations of Molecular Genetics

In the long chain of DNA, made of four different *nucleotides*, the links in the chain are the same for each nucleotide, and they consist of a sugar (called deoxyribose) linked to the next sugar by a phosphate. Each nucleotide also has a *base* attached on the side of the link: adenine, thymine, guanine, or cytosine. The four nucleotides are conveniently symbolized as A, T, G, and C, respectively.

On the basis of X-ray crystallographic analysis of the distances and angles between the atoms, Watson and Crick deciphered the general structure of DNA as a double helix, that is, two interwound strands. The key to its genetic function, however, is not its helicity: it is the complementarity of the bases in the two strands. A and T always pair with each other, as do G and C, holding the strands together by weak, readily reversible bonds. The sequence of the bases in one strand thus defines the sequence in the other, like a photographic negative and a positive.

Unlike most profound discoveries, which take some time to be accepted, the elegant structure of DNA was immediately and widely seen to provide a clear and simple explanation for the mystery of gene replication. In this process the two strands of DNA would peel apart in a short region, breaking the weak base-pairing bonds and thus exposing their bases (which had been buried in contact with each other in the double helix). Each strand could then serve as a template that guides the formation of a new complementary strand—that is, in the synthesis of the growing new chain each added nucleotide must be paired with the complementary base in the old strand. In this way the original DNA gives rise to two exact copies, each with one old and one new strand. It took many years before biochemists could accomplish this process in the test tube and analyze its complex enzymatic mechanism, but confidence in the fundamental mechanism, though based on very indirect evidence, did not waver.

Among the advances that followed identification of the structure, it was eventually recognized that each gene is not a definite molecule, as had always been assumed. It is a particular region, a few hundred to a few thousand nucleotides long, within the much longer chain of DNA (millions of nucleotides), like a sentence in a book. The enormous length of DNA had been concealed for many decades by the fact that its stiff double-stranded chain, unlike previously known molecules, is broken into fragments by the usual methods of isolation.

Distances along DNA are generally expressed in terms of number of *base pairs* (bp). The DNA of the *E. coli* chromosome amounts to about 4 million, while that of one haploid set of twenty-three human chromosomes is about 3 billion. Size can also be expressed in terms of physical length: the DNA in a set of human chromosomes, stretched out as a double helix, would measure about six feet.

The possibility of very strong selection for certain mutants in bacteria or bacteriophages radically improved the precision of mapping the chromosome. Classical geneticists, dealing with organisms such as fruit flies, could handle populations only of a few thousand, and they could only count, not select, the various phenotypes. Bacterial geneticists, however, could deal with populations of billions, and from these they could isolate very rare recombinants if they were highly selectable—for example, a drug-resistant recombinant that could grow, unlike the predominant population, in the presence of the drug. Since genetic mapping, as we have seen, depends on measuring the frequency of crossover between two sites on a chromosome, the new ability to detect exceedingly rare recombinants refined the resolving power of mapping by a factor of a million.

With this tool the sites of mutation, and also of recombination, no longer were mapped in terms of genes as the smallest units. Seymour Benzer, Charles Yanofsky, and Sydney Brenner showed that one could now map down to the level of individual nucleotides in the DNA; for while recombination had long been assumed, on the basis of classical genetic mapping, to occur only at special sites between genes, it was now seen to occur between any adjacent nucleotides. Moreover, mapping with such refinement soon demonstrated a linear correlation between DNA sequences and the corresponding protein sequences—a principle that soon permeated nearly all subsequent work in molecular genetics.[3]

Further studies revealed the mechanism of *gene expression:* the process by which the sequence in a gene, serving as a blueprint, determines a corresponding sequence in the many copies of its product, which become parts of the working machinery of a cell. There are two major steps, transcription and translation, in gene expression. In *transcription* the sequence of one strand of a gene is copied to yield a complementary sequence in *RNA*, which differs from DNA very slightly, in having the sugar ribose instead of deoxyribose. The RNA peels off the DNA as it is synthesized. Some of this RNA is a stable part of the working machinery of the cell. Much of it, however, is a less stable intermediate in gene expression, *messenger RNA*. The messenger sequence is then *translated*, on a complex apparatus of protein synthesis, into a corresponding sequence of amino acids, like a tape giving instructions in computer language to a computer that then prints out in English.

Translation is synonymous with protein synthesis, though one term emphasizes the genetic concept of information transfer and the other emphasizes the biochemical mechanisms. It takes place on the most complex of biochemical units, the *ribosome*, made up of fifty-five different molecules of protein plus several stable RNA molecules. The *genetic code* for the translation of messenger RNA sequence into protein sequence is virtually universal in all organisms. In this code each successive set of three nucleotides in the messenger RNA, constituting a *codon*, is specific for a given amino acid (or for "start" or "stop"), and it instructs the ribosome and its associated factors to insert that amino acid into the growing protein chain.

Each incorporation of an amino acid involves many successive steps and a large expenditure of energy—the price paid for synthesis at a high rate, with great accuracy, and with unlimited choice of sequence. This large expenditure of energy has another consequence, noted above in the discussion of Lamarckism: it makes the flow of information from nucleic acid to protein sequence irreversible.

Methodological Advances and Practical Applications

Despite these and many other profound advances in our understanding of cellular processes, the practical payoffs of molecular genetics were awaited for decades, with increasing impatience. There were good scientific reasons for the delay. The molecules identified by earlier biochemical studies, such as amino acids, vitamins, and hormones, could be immediately put to use because they were small enough to be able to penetrate, like drugs, to locations where they had the desired effect. In contrast, the much larger molecules of nucleic acids and proteins, on which molecular genetics is based, cannot enter cells from the bloodstream but must be made in the cells that use them.

A new breakthrough was needed, and it finally arrived in the form of molecular recombination of DNA. Among its many contributions, this tool has made it possible to identify a large number of genes specifically involved in cancer (oncogenes), and to dissect their functions in normal regulation of cell growth as well as in the unregulated growth of cancers. While this knowledge has still not been translated into clinical practice, the smell of success is in the air, and the field has attracted many of the brightest minds in biomedical research.

In the development of molecular recombination the key discoveries were the following.

First, many bacteria make various enzymes, called *restriction endonucleases*, each of which cleaves DNA at a specific short sequence (a *restriction sequence*). In nature this action damages those viruses whose DNA has that sequence, thus restricting the range of hosts that they can infect. (The host cell protects itself from its own endonuclease by enzymatically modifying bases in the key sequence, in a way that does not alter their coding function. There is evidence that these endonucleases, though discovered as agents of restriction, have a deeper function, promoting recombination and hence evolutionary variation.)

Second, most restriction enzymes do not cleave the two strands of a restriction sequence at the same position. Instead, these sequences are palindromes, and the endonuclease cleaves them symmetrically at sites four to six nucleotides apart, thus creating an "overlap" between the new ends. When these ends then separate, each has a short single-stranded "shelf" extending beyond the double-stranded DNA.

Third, these single-stranded ends, having been paired to each other before separation, have complementary sequences. As a consequence, they are "sticky": that is, under conditions favorable to pairing rather than to separation each can pair with a complementary end, on any fragment made by the same enzyme, from any DNA. This pairing involves the same weak bonds that normally hold two DNA strands together in a long double helix.

Fourth, an enzyme called a *ligase* fuses the temporarily paired fragments into stable double-stranded DNA by forming a strong bond in each strand, at the joint between the two fragments. The entire process is often called *splicing*.

And fifth, the recombinant DNA can be taken up by bacteria under appropriate conditions. If it is a fusion of chromosomal fragments, lacking the genes required for self-replication, it must be inserted into the host chromosome, by recombination, in order to be perpetuated. Alternatively, if a fragment has been incorporated into a self-replicating *vector* it may replicate as part of that vector. There are two kinds of vectors: nonvirulent bacterial viruses, which can replicate in a host cell without killing it; and *plasmids,* blocks of DNA that also can replicate in the cell separately from the host chromosome, but that do not form a viral coat to assist spread to other cells. Any of these methods of perpetuation results in *cloning* of the recombinant DNA, that is, formation of multiple identical copies as the cells divide.

Since all these steps, and the enzymes that investigators use to accomplish them, occur in nature, there can be no doubt that foreign DNA enters bacteria in nature by the same mechanisms, but at a very low rate. Investigators have not created the several steps: they have only modified them, in ways that make them more useful.

The technology of cloning recombinant DNA made it possible to obtain unlimited quantities of any gene that has been isolated. It also provided a radically new way of obtaining specific proteins in quantity, because genes can form their product in a foreign host cell if also provided with regulatory signals in the DNA that that cell can recognize.

Moreover, starting with bacteria, the range of hosts has been extended to yeast, animal, and plant cells, and more recently to whole animals and plants.

Other methodological advances have further enhanced the powers created by the recombinant DNA technique. These include simplified methods for purifying proteins and cloned DNA fragments; rapid methods for determining their sequences, and for synthesizing desired sequences; and *directed, site-specific mutations*, achieved by either genetic recombination or synthesis of DNA, to yield predictable novel proteins. In a particularly bold, rather early technique, called *hybridization*, DNA is incubated at temperatures high enough to separate the strands, and on gradual cooling (annealing), complementary fragments frequently pair with each other to form double-stranded DNA. This procedure can be used to identify the location of a specific DNA sequence, on a chromosome or in a set of separated fragments, by hybridizing it with a radioactively labeled known segment of DNA. Similar hybridization with the messenger RNA in a cell can be used to determine which of the genes in that cell are being expressed.

A particularly powerful technique, called the *polymerase chain reaction*, is based on a special kind of enzymatic copying of DNA, which is initiated by strand separation and hybridization of each strand, at opposite ends of a desired region, with a short sequence of RNA acting as *primer* for DNA replication. The two strands of DNA are copied in opposite directions, starting at the primer, thus yielding two double-stranded copies of the region of interest. By repeated cycles of strand separation by heating, then cooling, priming, and copying, the desired segment of DNA is multiplied exponentially (like multiplying bacteria), rather than at a constant rate. As a result, even a single molecule of DNA—say from the dried tissue of a mummy, or in a cell attached to a hair root that might identify a criminal—can be multiplied rapidly to yield a large amount. This technique is replacing cloning for many purposes.

Molecular genetics is revealing an expanding number of protein molecules that play specific roles in health and disease, including various enzymes, regulatory proteins, and the surface receptors that mediate interactions of a cell with its environment. Meanwhile, physical methods have been improving for determining the three-dimensional structure (i.e., the spatial arrangement of the atoms) of proteins. These two approaches together are offering increasing promise of a more rational search for useful drugs, by guiding the synthesis of small molecules that will fit the precise shape of a site directly involved in the activity of a protein.

Finally, computer programs have been developed for scanning files of long DNA sequences to identify degrees of similarity between different genes within an organism, or between homologous genes from different organisms. These similarities are important not only for tracing the branching, and its timing, in the evolutionary tree, but also for tracking down the function of a gene—for example, the known function of a gene in yeast or the mouse can help to identify the function of a related sequence in humans.

Conceptual Contributions of Molecular Genetics

In addition to advancing our technical powers, molecular genetics has contributed to a number of developments that might broadly be called philosophical. One effect is seen in the rapidly changing intellectual organization of the life sciences. For most of their history the branches concerned with origins (evolution, genetics) have been quite separate from those concerned with function (physiology, biochemistry, developmental biology), but now practically every branch of biology and medicine uses the same molecular genetic concepts and techniques in its research.

In a more profound contribution, molecular genetics has finally provided a rigorous, marvelously simple answer to one of the fundamental questions in biology: What distinguishes living from nonliving matter? Features proposed in the past, such as catalysis, production of energy, and contractility, have their inorganic counterparts. But molecular genetics has now revealed a unique feature of the organization of matter in living organisms: the storage and transmission of information in giant molecules.

This concept of molecular information arose from studies on the function of the nucleic acids. In these molecules the information is encoded in variations in a one-dimensional record, much as in writing or a computer tape; and it is transmitted either by being copied into a complementary nucleic acid sequence, or by being translated into a sequence of amino acids in the synthesis of proteins. But in addition to this form of information storage there is another, less widely recognized form of intramolecular information, which is not a stable repository but changes rapidly in response to changes in the current environment. This information is located in the three-dimensional shape of a special class of regulatory proteins, in the following way.

All proteins are synthesized as chains of amino acids of twenty different kinds, in a specific sequence from a few dozen to a few hundred units in length. The specific affinities of various amino acids for each other, involving weak bonds like those between the strands in DNA (but with greater spatial flexibility), then cause the one-dimensional chain (called a *polypeptide*) to fold into a stable, specific three-dimensional shape, with a unique surface pattern that determines its interactions. However, certain regulatory proteins have evolved to have not one stable state but two, one being active and the other inactive in the regulatory function. Such *allosteric* (alternative shape) proteins are shifted reversibly by binding or releasing a regulatory metabolite. Since the proportion of the molecules of the protein present in either state depends on the concentration of its regulatory metabolite, the protein serves to sense that concentration.

This information about metabolite concentrations regulates the internal economy of the cell by adjusting the rates of synthesis, through a feedback mechanism, to maintain each metabolite within a narrow, appropriate concentration range, much as a thermostat maintains the temperature of a house. The thousands of compounds in a cell are thus economically synthesized, in proper proportions. Allosteric regulation also guides interactions of the cell with the external environment, such as induction of enzymes to utilize a new food or movement of bacteria toward a higher concentration of that food.

There are two major classes of regulatory proteins. Generally, the first enzyme in a particular pathway senses the end product, thus serving like a valve at the beginning of a branch in a pipeline sensitive to pressure at the end. Another class of allosteric proteins are pure regulators and not enzymes, and they bind to appropriate targets in cells, such as DNA sites that govern the activity of specific genes, or proteins that contract and thus cause movement.

Allostery was discovered as an information feedback system in bacteria, where a metabolite provided by the external environment turns off its own synthesis in the cell, thus sparing energy and material for other reactions. In addition, the same feedback mechanism also regulates the internal economy, ensuring that each metabolite is synthesized at a rate that meets the cell's needs but does not result in wasteful spillage to the exterior. Since both these contributions to metabolic economy increase the growth rate, it is not surprising, in retrospect, that such primitive organisms as bacteria have already evolved such elaborate regulatory mechanisms: the slightest difference in growth rate of two strains will have a profound effect, over evolutionary periods of time, on their relative success.

The allosteric mechanism was soon found to have much broader biological significance than merely increasing the growth rate of microbes. Not only do allosteric proteins regulate metabolite levels in all cells, but they respond to a wide variety of other stimuli, within and between cells. These include hormones, most drugs, and the function of the nervous system. Moreover, the expanding study of molecular regulation further showed that information is also encoded in certain proteins through more persistent, enzyme-catalyzed modifications that change their shape, such as the addition or removal of a phosphate group. This kind of change is very important in transmission and storage of information in the nervous system, as will be shown in the chapter on that topic. Who could have imagined that a principle of such broad significance would have emerged adventitiously from studies of growth rates in bacteria?

Allosteric regulation has also had an impact on an ancient principle in biological thought, *teleology* (Greek *tele*, goal). This is the assumption that a structure or function would not be present in an organism unless it had a purpose or goal. The original, Aristotelian version of this idea assumed that the specific goal had somehow been foreseen. However, in the modern version, called *teleonomy*, the purpose or usefulness of a feature has resulted not from the foresight of a creator but from the hindsight of natural selection. For decades many scientists considered teleonomic arguments too speculative and insisted that descriptions of mechanisms should not be accompanied by inferences about their value for the organism. Now, however, allostery has made the concept of teleonomy thoroughly respectable: regulation would have no meaning unless linked to the concept of a goal.

In addition to its implications for biology, the discovery of two different mechanisms of information transfer, via nucleic acids or via allosteric proteins, has epistemological implications. It provides a concrete foundation for an obvious but not universally accepted notion about the nature of our knowledge of the external world: that it is of two interacting kinds, inherited and learned. Inherited knowledge is programed in our genes; in the developing organism it is expressed in the organization of the emerging nervous system; and the result is a brain with capacities and instincts, mostly common to the species but partly unique to the individual. In contrast, evidence is accumulating that learned knowledge, acquired by experience, is recorded in modifications in the number or the functional properties of synapses (the connections between nerve cells) in our brain; and these modifications must involve allosteric proteins.

This insight from modern biology fits Kant's view that we start our lives with a great deal of a priori information, to which we then add learned information. But while our inborn information is a priori for individuals, we now recognize it as a posteriori for the species, as a product of evolution. If evolution had not provided an organism with a good deal of inborn information that influenced its responses to stimuli, and if these responses did not correspond reasonably well to the realities of the external world, our predecessors would not have survived—and they could not then have evolved into creatures that can argue about epistemology.

One more philosophical implication of the success of the genetic revolution is its encouragement of reductionism, in one sense of the term, as an approach to science: the proposition that when we study a phenomenon we increase our understanding if we can reduce it to events at a lower or finer level of organization of nature. It is important to distinguish this view from another, belittling meaning often ascribed to the term *reductionism:* that only studies at the finest levels are important or fundamental. On the contrary, we can learn a great deal, for example, about behavior at the psychological or the neurobiological level without knowing its molecular basis—though an integration of the multiple levels is the ultimate goal. Moreover, interpretations at an abstract, inferred level gain in interest as they can be integrated with events closer to our ordinary perceptions. However, the connections between levels cannot be forced or predicted. In Lederberg's wry question, who would have guessed that we would understand the molecular structure of genes before that of tendons?

Earlier biologists sought novel "vital forces" to explain life, and more recently some physicists sought novel physical laws. But molecular genetics has explained a major mystery, the action of the genes, without such hypotheses. The concept of information storage in nucleic acids or allosteric proteins employs only standard physical forces and cannot be considered a new physical law, though it is a new principle in biochemistry.

The same reductionist, analytical approach promises enormous further increases in our understanding of biological processes in the decades and centuries to come. At the same time, we must also recognize the narrow limits of this approach. For despite its spectacular success in answering questions about the nature of the external world, science does not provide objective answers to another set of questions, of paramount importance in our daily lives: those involving values. A related question, whether we will be able to explain consciousness in materialist, neurobiological terms, I discuss in Chapter 14.

Notes

1. Retrospectively, it is tempting to suggest that if evolutionary biology had been launched after the development of genetics, rather than long before, it might have generated less resistance. See B. D. Davis, "Molecular Genetics and the Foundations of Evolution," *Perspect. Biol. Med.* 28 (1985): 251.

2. Avery's finding had the two elements of a great discovery: surprise and broad significance. Yet he did not receive a Nobel Prize, though he lived for

eleven years after announcing the role of DNA. For a more detailed discussion see B. D. Davis, "Molecular Genetics, Microbiology, and Prehistory," *BioEssays* 9 (1988): 129.

3.   Like Avery's discovery, this development of fine-structure genetics did not receive a Nobel Prize, though it involved both surprise and broad significance. Perhaps its significance was not fully recognized because it so quickly became taken for granted as a foundation on which so many others built, within an already flourishing field.

3

Microbes: The Laboratory and the Field

Allan M. Campbell

Shortly after World War II, research units of the U.S. Army were study-ing ways to wage or defend against biological warfare. One favorite method for delivering disease germs effectively to their targets was to generate an aerosol containing many bacteria dispersed in a mist of liq-uid droplets that could be sprayed from the air. Many details of the oper-ation are still secret, but someone in authority agreed to evaluate the efficiency of delivery by actually spraying large numbers of living bacte-ria over populated areas. Several U.S. cities, including San Francisco, were chosen as targets. The Army microbiologists chose as a test organ-ism the bacterium *Serratia marcescens.* This bacterium combined the virtues of harmlessness to human beings (as could be verified by con-sulting the bacteriology textbooks of the time) and easy identification (because the colonies formed on culture plates are blood red). During 1950–51, large quantities of *Serratia marcescens* were sprayed over San Francisco, and the efficiency of dispersal was monitored by exposing Petri dishes in various locations nearby.

It is now appreciated that many usually harmless bacteria, nor-mally present in the environment, can cause infections in individuals whose defenses have been weakened by various defects. In San Fran-cisco several unusual infections with *Serratia marcescens* were en-countered in hospitals at the time of the spraying by the Army. When the secrecy was lifted twenty-five years later, the possibility was raised that some of these infections might have been caused by the bacteria released by the Army. However, in the one fatal infection that was the object of legal action, a different strain of Serratia seems to have been responsible. It therefore remains undemonstrated that the Army test had caused any damage at all.[1] Nevertheless, the episode raises the question whether massive exposure of human populations to any kind of bacteria should be regarded as totally safe.

A more recent introduction of bacteria into the environment was proposed by agriculturalists rather than the military; and though it built on a long tradition of using bacteria in various ways to promote growth of plant crops, it aroused much concern because it involved a genetically engineered rather than a naturally occurring bacterium. During the 1970s plant pathologists discovered that the formation of frost on crop plants is facilitated by the presence on leaves of certain bacteria—including members of the genus Pseudomonas—which act as nuclei around which the ice crystals form. The mechanism of nucleation is not fully understood, but it was found in the 1980s that mutational inactivation of a single gene of *Pseudomonas syringae* greatly diminished its nucleating ability. Furthermore, if a large number of the non-nucleating mutant bacteria (or of similar naturally occurring strains of Pseudomonas) were added to a plant along with some nucleating bacteria, the non-nucleating bacteria could, by simple competition for resources, reduce the multiplication of nucleating bacteria, thereby protecting the plant from frost damage. For the practical purpose of applying such bacteria to crop plants in the field, the competitor chosen was neither a natural strain nor an ordinary laboratory mutant, but rather a mutant strain artificially constructed by genetic engineering. The normal gene, called *ice+*, was isolated, chemically altered in the test tube so as to delete much of it, and then reintroduced into bacteria, and finally bacteria were isolated in which the altered gene, called *ice−*, had replaced the normal one. These operations produced bacteria that were changed as little as possible from their natural counterparts except for the alteration of this one gene, and that gene was so extensively damaged that additional mutations could not restore the lost information.[2]

Because the bacteria had been produced by artificial genetic manipulation, their inventors had to seek approval for their introduction into the field from the Recombinant DNA Advisory Committee (RAC) of the National Institutes of Health.[3] The RAC met, deliberated, and approved field testing of the Ice− bacteria. Two considerations were paramount in reaching this decision. First, the Ice− bacteria constituted no apparent hazard. Second, deletion mutations equivalent to those that had been engineered are expected to occur in natural populations and therefore cannot constitute biological novelties. The decision of the RAC was later challenged and led to extensive litigation.

These two episodes illustrate the basis for a conclusion that has been reached whenever the regulation of genetically engineered organisms has been analyzed scientifically: for regulatory purposes these organisms should not be distinguished from natural organisms. Harm may or may not result from the dissemination of some genetic engineered organisms, and the potential for harm may not always be obvious in advance; but, as with the *Serratia marcescens* over San Francisco, the same can be said of natural organisms. Wherever a valid cause for concern is evident, the concern is based on the properties of the organisms used, not on their mode of origin.[4]

Despite the logic of this conclusion, governmental regulations in most countries will probably continue to distinguish engineered organisms from natural ones for the foreseeable future. This fact, and the scientific discussions of the issues, are best appreciated in the context of the historical and legal background, starting with the remarkable course of events following the discovery of artificial gene splicing.

The Safety of Laboratory Experimentation

The techniques developed in the early 1970s for splicing DNA segments in the test tube and reintroducing them into living cells made possible in principle the deliberate creation of genetic novelty on a scale that had not previously been imaginable. Almost before the experiments had progressed beyond the drawing board, some molecular biologists initiated efforts to delay their application.

The psychological and sociological factors that promoted these efforts will not be discussed in detail here.[5] Certainly the general level of political activism that prevailed in the wake of the Vietnam era had an important influence. The populist doctrines of participatory democracy attractive to some scientists of the New Left doubtless affected the manner in which the subject was presented to the public, as did the reluctance of other scientists to thwart such doctrines. Some scientists expressed concern about specific dangers from the new methods, which might allow production of harmful human viruses on a much larger scale than would otherwise be possible. The major concern was the scale, not the novelty of the method. These methods might also be used to transfer genes for antibiotic resistance into disease-causing bacteria, thereby rendering antibiotic therapy ineffective.

A committee of eminent scientists convened by the National Academy of Sciences and chaired by molecular biologist Paul Berg distributed a letter, sometimes referred to as the "Berg letter," calling for a voluntary moratorium on the cloning of genes for antibiotic resistance and of DNA from animal viruses pending an international conference to discuss the issues. That conference was held at Asilomar, California, in 1975. The atmosphere at the conference has been most aptly likened to that of a religious revival meeting.[6] Many (not all) of the scientists who spoke were swept into the role of the Moral Scientist who places the Public Good above Individual Ambition. There was little serious discussion of what really constitutes the Public Good. Though dubious, the assumption that the meeting had been called to address a serious problem, rather than to determine whether any problem existed, went largely unchallenged.

The Asilomar conference set in motion the political machinery that led to the establishment of the RAC. One of the first acts of the RAC was to produce a set of Guidelines for Recombinant DNA Research, to be followed by all institutions that received any support from the NIH and eventually by most other government agencies as well. Scientists gradually mastered the euphemisms of government. "Guidelines" were not supposed to provide guidance to the intelligent user but rather constituted regulations to be obeyed line by line, even though the scientists who had drafted them intended otherwise. The RAC was indeed "advisory," but not to the investigators; rather, it advised the director of the NIH, who then converted its advice into mandatory rules.

Experiments were classified into a complex hierarchy of risk levels, but the classification lacked scientific logic. The greater the *perceived* level of concern that had been expressed over certain types of experiments, the higher the assigned risk level. This process was roughly equivalent to comparing the relative health hazards of cooking with microwave ovens and of nude sunbathing by taking a public opinion poll to see what fraction of the population regards each of these activities as dangerous.

The first edition of the guidelines—which ultimately went through numerous revisions—paid inadequate attention to the role of competition in limiting the spread of microorganisms. Certainly bacteria can replicate very rapidly, and the phages and plasmids used as cloning vectors in recombinant DNA research can spread rapidly from one bacterium to another. But the general principles of natural population growth, which have been known since Darwin and Malthus, imply that the potential rates of spread observed in the laboratory will almost never be realized by a few stray bacteria that happen to escape from the laboratory into an environment that is already filled with natural bacteria.

To illustrate the point, imagine a naive horticulturist who has succeeded in breeding a pink dandelion and awakens during the night to realize, "If I plant one seed from my dandelion in a field, that one plant may make a thousand seeds, and next year there will be a thousand pink dandelions. Each of those plants will make a thousand seeds, and a million pink dandelions will result. In a few years the whole earth will be carpeted with pink dandelions." Obviously this will not happen. Every yellow dandelion in the field is already producing a thousand seeds—or whatever number a dandelion actually makes—and each of these thousand seeds has, under natural conditions, one chance in a thousand of surviving to form a mature plant. Assume that the same numbers apply to the pink dandelion, that is, the fitness of the pink dandelion is the same as that of yellow dandelions. Then, if anyone sets out one pink dandelion and comes back ten years later it will, on average, have exactly one living descendant and no more, for the living world has long since become saturated with dandelions. Bacterial habitats are likewise saturated with bacteria, which in turn are saturated with phages and plasmids that infect them. Natural populations are seldom approaching the Malthusian limits of saturation density; they have already attained them, and on a global scale the expansion of some local populations is balanced by the contraction of others.

To say that this balance is the rule is not to deny the occasional exception, such as the dramatic temporary increase in numbers of disease-causing bacteria during an epidemic (which depends on a favorable environment, namely, a high concentration of susceptible, nonimmune hosts). But any policy that ignores the rule is grossly flawed. For example, one construction that was once treated as virtually a blueprint for disaster is to clone a potent toxin gene such as botulism toxin into a natural inhabitant of the human gut such as *Escherichia coli*; and even worse, to put the toxin gene on a plasmid capable of causing its own rapid transfer to other bacteria of the gut. Whereas there is some prudence to handling such a bacterium with respect, it is very doubtful that the escape of a few such bacteria into the environment would have any impact at all, unless the presence of the toxin gene conferred some selective advantage to the bacterium or to the plasmid. It is probable that all the toxin genes have gotten into plasmids at some time or other and have gained access to exotic cells, but the failure to find such recombinants in nature indicates that they have no selective advantage.

The first edition of the NIH guidelines forbade the cloning of toxin genes. Even experiments that might permit their accidental cloning were allowed only in elaborate facilities available to few if any investigators. The concern underlying these extraordinary restrictions was the possible creation of new disease-causing bacteria that might spread through the human population. In some diseases, such as cholera or diphtheria, a single potent toxin is responsible for the major disease symptoms. But even there, toxin production is just one of the numerous genetic characteristics required to make the bacterium an effective pathogen. The ability to invade and colonize a human host as well as to spread from one host individual to another is a complex, highly evolved property. The payload (toxin) has little impact without an effective delivery system; and that delivery system requires a constellation of specific characteristics highly unlikely to arise by chance. It would be an exaggeration to claim that we know all of the possible routes whereby a bacterium might become pathogenic; but the results of conventional genetic crosses between pathogens and nonpathogens all indicate that multiple changes are required to change a nonpathogen into a pathogen.

Escape from small-scale laboratory experiments can be of concern only if the engineered organism is selectively superior to its natural counterparts. Neither for pathogens nor for nonpathogens do scientists know how to engineer such selective superiority. To go back to our pink dandelion, it might be asked what would happen if the pink dandelion should make two thousand seeds, as compared with the one thousand made by its wild counterparts. Would it then in fact overgrow the natural population and change the face of the earth? The answer is that it probably would—if it were possible to make those extra one thousand seeds at no additional cost to the plant, such as diversion of resources from other properties that promote survival, like leaf size or rapid germination. But if that were possible, the chances are enormously high that the natural plants would already be making two thousand seeds. Organisms that survive in nature are continually tested for optimization of reproductive strategies. That is not to say that all organisms have reached any ultimate optimum, but only that, at present, no one knows how to improve fitness beyond what nature has already achieved. Random changes may sometimes be selectively neutral but are frequently deleterious. The most reasonable expectation is that a pink dandelion would be less fit than its natural relatives, because flower color is itself a selected trait.

In 1979 a conference in Wye, Kent, led to the publication of a volume in which possible ways to arrive at a more sensible regulatory system were discussed not only by molecular biologists but also by experts in other relevant areas, such as the general problem of estimation of risks in industrial operation.[7] An instructive exchange took place between DeWitt Stetten (retiring chairman of the RAC) and Sir Frederick Warner, an expert on risk assessment:

> *H. D. Stetten:* Sir Frederick . . . referred to . . . carcinogenesis, teratogenesis and mutagenesis as the three major anxieties. Each of these in a favourable situation can be approached, I think, in an actuarial fashion, and one has ample data on which to base a

risk assessment. Only when the increment in hazard is small compared to the noise, as Sir Frederick pointed out, do we start to have real problems. But, with recombinant DNA, we have a situation which is different. The first mouse, as far as I am aware, has yet to be killed; the first human has yet to be damaged by this technique. Can you give us any help as to how we assess risk when we have zero actuarial basis?

F. Warner: There is no help available, and somebody has said you got yourself into this position and I think you have to live with it. Once you have sounded the alarm about risk then you have put into people's minds the fact that there is a risk. You did that presumably on the basis that one was suspected without perhaps taking time to get together all the evidence which there might have been from related fields to see if you could put into perspective what an actual risk was.

Whenever you have talked about risk assessment you really have been trying to decide whether there is, in fact, a risk at all, whereas in every other field where we talk about risk assessment, we have the historical data, we know the risks, we have gone in with our eyes open . . . You go into engineering as you go into any form of economic activity, and if you spend a million pounds you are going to kill somebody. I am sorry this is a brute fact of life.

These comments tell us that the attempts to identify hazards specific to laboratory experimentation with recombinant DNA are a chase after phantoms. In all the discussions since 1975, no evidence had surfaced to indicate that recombinant DNA was dangerous; and if it was not dangerous, there was neither a need for regulations nor any logical way to draft them. The comments have two other implications as well. First, since no activity is totally risk-free, appreciable hazards will always surface when the scale is sufficiently large. Second, the wisdom of addressing conjectural risks by calling for moratoria and organizing highly publicized conferences is questionable.

The Scale of Operation

When the use, production, or dissemination of known hazardous agents is legally regulated, the law usually makes some distinction of scale. Thus, prohibitions against the discharge of toxic wastes into sewers specify maximum permissible amounts; and Occupational Safety and Health Administration (OSHA) regulations requiring special precautions in handling concentrated solutions do not apply to dilute solutions. In contrast, legislation that restricts agents without regard to concentration, such as the Delaney Amendment, which prohibits adding cancer-causing agents to food, is generally impractical to administer and leads to regulatory overkill.

In a sense, the use of bacteria that cause human disease is also regulated according to scale. The Centers for Disease Control (CDC) classify infectious agents on a graded scale determined by severity of infection, ability to spread, and rarity within the United States. Class 1 agents are harmless, Class 4 perilous. Class 2 includes agents such as bacteria of the genus Salmonella, some of which cause typhoid fever and food poisoning. Deliberate contamination of the environment by Salmonella is forbidden by the CDC guidelines, which have no direct force of law but are legally accepted as standards of good practice. However, the stringency of the containment measures required for laboratory use of Salmonella is much lower than that for a Class 4 agent like Lassa fever virus, a virus of African rodents that is fatal when transmitted to susceptible humans. The CDC guidelines thus allow a low probability of escape of Salmonella from laboratories. If this low probability per usage is multiplied by a large number of hours of laboratory experimentation, some escape will doubtless occur. Thus, the CDC guidelines in effect authorize a low rate of environmental contamination by Salmonella escaping from research laboratories. In the same sense, the NIH guidelines authorize a low rate of escape of laboratory DNA recombinants.

Scale can relate to hazard in two ways. First, small unit risks may be additive as exposure is increased. For one person lying on the beach for one afternoon, there is only a tiny chance that the ultraviolet irradiation absorbed will cause skin cancers, but among a million sunbathers, some will be affected. Second, there can be threshold effects, reached only at high doses of deleterious agents. Swallowing a cup of concentrated acid can ruin the digestive tract, whereas the same amount of acid consumed in small doses may be altogether harmless. Or small amounts of some substances, such as lead, may accumulate and exceed a toxic threshold.

For several years after their introduction, the NIH guidelines prohibited altogether any large-scale propagation of genetically engineered microorganisms, "large-scale" being arbitrarily defined as any volume above 10 liters. At that time, the major concern was with accidental escape of even a single recombinant bacterium. As far as I know, the prohibition against large amounts derived from the notion that they might be difficult to contain, not that escape would have more serious consequences. Increased appreciation of the role of natural competition in limiting the spread of bacteria, once they have escaped, led eventually to the exemption from the NIH guidelines of all gene cloning performed in certain laboratory strains of microorganisms that are demonstrably at an extreme competitive disadvantage relative to their natural counterparts. Laboratory strains, like hothouse plants, seldom retain the ability to compete well in the wild, and it seems extremely unlikely that their engineered derivatives will fare any better.

If small-scale experimentation, with the possibility of some escape, is thus exempted, why should large-scale experimentation, industrial-scale production, or even deliberate introductions be treated differently? Even if the strains used are not deliberately disabled, the engineered properties are unlikely to give them a competitive *advantage* in the real world. But do not forget Sir Frederick's admonition, "If you spend a million pounds you are going to kill somebody." Almost any technology, applied extensively enough, will have some adverse fallout, which it is desirable to anticipate and minimize. Planned introduction to a garden plot is not the same as spraying millions of acres of crop plants, but it is one step in that direction.

With minor exceptions, the rules of competition do not change with scale. If a ton of Ice$^-$ bacteria are broadcast over a field, the bacteria are not expected to invade neighboring territories and take over the world, any more than would the descendants of a single bacterium.[8] But the possibility that the large amount of bacteria might itself cause some harm cannot be dismissed summarily. The bacteria, for example, might have a very small unit probability of affecting human health, as did the *Serratia marcescens* over San Francisco. Or, in large numbers, they could cooperate to exert some effect. It has been suggested, for example, that Ice$^+$ bacteria help to seed clouds, and that replacing them with Ice$^-$ bacteria over a large area might have adverse meteorological consequences. Both these possibilities may be unlikely, but it is legitimate to raise them as concerns and to expect some reasonable effort at evaluation.

The legitimacy of these concerns in no way discredits the decision to allow field testing of Ice$^-$ bacteria, or the position that engineered organisms require no special regulation. The same concerns would exist if the testing were being conducted with natural Ice$^-$ bacteria or with nonengineered Ice$^-$ mutants (in which case no prior approval for field testing would have been required). This was precisely the basis of the RAC's decision. Its primary responsibility is to give expert advice on any hazards that might have been created by genetic engineering. The ordinary hazards that might attend dissemination of natural bacteria are beyond its jurisdiction. And the position that engineered organisms require no special regulation implies that whatever the agency with jurisdiction—including in principle the Environmental Protection Agency, OSHA, the Food and Drug Administration (FDA), and other federal agencies, each with its own mandate—it should apply uniform criteria to engineered and natural bacteria.

Had the RAC perceived any actual hazard in the operation, even one unconnected with genetic engineering, it would doubtless have considered itself responsible to report its concerns so that the appropriate agency might act. The RAC is in no sense passing the buck by remaining within the scope of its charge. Its proper role in the review process touches in several ways on the relationship among scientific judgment, public opinion, and legal decision making.

Science and Public Policy

The RAC operates on premises that most scientists accept without reservation. When a new activity is undertaken by individuals or organizations, the community at large has a proper interest in finding whether the activity creates hazards to public health, environmental quality, or anything else that affects the general welfare. What better way to evaluate the potential hazards than to convene a committee that includes scientists with sophisticated technical knowledge, plus a few informed nonscientists to keep the scientists honest? In areas where the risks, if any, must be specific to the particular materials and methods employed, isn't that how everyone would want to proceed, if it were possible? If everyone agrees that dangerous activities should be restricted and that harmless activities should proceed with minimal interference, isn't the best regulatory system one that discriminates between them in the most informed and intelligent manner possible?

These questions may seem rhetorical, leaving room for doubt only as to whether a procedure such as RAC review meets the intended goals. However, there are genuine and honest differences about ends as well as means. One complication is that the legal system traditionally accords enormous respect to precedent. A scientist may consider it totally illogical to prohibit one activity and allow another when the inherent risks are identical, but the law does so routinely; and, in some measure, the average citizen expects it to. Pharmaceutical companies and their scientific advisers complain bitterly that if aspirin were a new drug, current FDA procedures would probably prevent its entrance into the marketplace. A new product must meet criteria more stringent than those applied to products already in use. A product can be removed from the marketplace on evidence of demonstrable harm, but the burden of proof is shifted. Society in general supports such a double standard. Many people might agree to the proposition that marijuana and ethanol are equally injurious to human health, but far fewer would vote to legalize the first or prohibit the second.

If everyone accepts the principle that new activities may be regulated on criteria somewhat more stringent than those applied to existing activities, where does that place genetically engineered organisms? Clearly each engineered organism is to some extent "new," or the engineering would be unnecessary. Does approval of one engineered organism, however benign its status, set a legal precedent that somehow shifts the burden of proof with respect to other engineered organisms? Some of the most vocal opposition to the Ice⁻ release is apparently based on the assumption that it does.[9]

The scientific judgment that there is no basis for distinguishing engineered from natural organisms has important corollaries. One is that however the law may choose to distinguish between new and established practices, the widespread dispersal of any organism should be treated in the same manner, whether or not it is engineered. This goal could be achieved by imposing either more regulation on natural organisms or less on engineered ones.

Another corollary is that there can be no generic tests for the safety of engineered organisms as a group, because engineered organisms as a group cannot be distinguished from natural ones by their properties. The various generic arguments that have been raised so far for either the safety or the danger of engineered organisms can be dismissed as either unproved or illogical.[10] The "safety" arguments include, "Engineering can never improve fitness, and therefore engineered organisms will never spread very far on release." This argument assumes that all gene combinations creatable in the laboratory have in fact arisen at one time or another in nature, and that they must have been discarded by natural selection. But in fact, only a minute fraction of such combinations can possibly have come up during the last 3 billion years. One can by the same token reject the corresponding "danger" argument: that genetic engineering will be more likely to improve fitness accidentally than the natural genetic lottery, which is constantly creating novel combinations of unknown properties by recombination and mutation.

These conclusions may seem obvious, but they have not always been followed in practice. In the late 1970s, when the regulation of recombinant DNA experimentation was at its peak, it was decided that it might clear the air if a test were made of one scenario about which concern had been expressed. Was there any chance at all that cloning the DNA of an animal tumor virus into a bacterium that normally resides in the human intestine might produce an epidemic of human cancer if the bacterium escaped from the laboratory? Millions of dollars in public funds and many investigator-years were consumed in demonstrating that a disabled laboratory strain of E. coli carrying the DNA of the polyoma tumor virus does not transmit the virus to the mouse in whose intestine it grows.[11]

The experiment served a useful purpose in politics and public relations, probably helping to bring about some relaxation of the NIH guidelines. This was experimental "evidence," indicating that genetic engineering might not be so dangerous after all. But it did not really prove anything about the safety of genetically engineered organisms as a group. This particular scenario required a sequence of improbable steps. Even in the unlikely circumstance that some viral DNA made its way from the intestine to the susceptible cells in the mouse, it would need to enter those cells (another extremely improbable event); and even if all this happened, the bacterium would not cause an epidemic unless it could outcompete the indigenous intestinal bacteria and spread in the human population. Any scientist requesting funds to support research that presupposed such a juxtaposition of improbable circumstances would probably have been told by a peer review committee to try something more sensible.

If most other virus-in-bacterium constructs are analyzed for expected danger by such a priori considerations, the expectation is that they likewise will be safe. But the polyoma–E. coli result does not make that expectation any more secure. Anyone seriously worried about the possibility that genetic engineering might generate unknown or unpredictable hazards could hardly be swayed by the fact that one system had proved safe, as predicted. To be sure, a scientist must accept experiment as the ultimate source of information: but we can better evaluate the expectations for a heterogenous group of unknown objects by extending principles established for known objects, rather than by sampling one or two of the unknown objects.

One misimpression that has been created, to some extent deliberately, is that the relaxation of the guidelines in the early 1980s was based on the accumulation of solid and convincing evidence against the dangers feared in 1975. Actually there was little new evidence, and its relevance to the expressed fears was marginal. More important was the increasing realization of the shallowness of thought that had led to the early restrictiveness. It is misleading to label the whole controversy as a dead issue if "dead" is taken to imply that the concerns were ever scientifically viable.

Currently there is a new round of safety testing on microorganisms proposed for deliberate release, from Ice⁻ bacteria to genetically modified viruses for the control of insect pests. One view of such testing was given by Edward Adelberg: "General principles are extremely important. At some point we must rely on scientific principles to tell us whether we have enough data. Then, if the experiments suggest that most GEMs [genetically engineered microbes] won't compete or won't do harm, the burden of proof is on the opponents of deliberate release to produce plausible scenarios of harm."[12]

This statement may represent the most effective strategy for countering many objections to planned introductions, by implying that concessions to demands for safety tests should be limited in number. If taken literally, however, the statement may foster misimpressions. It seems to suggest that if Ice$^-$ bacteria prove safe, we should feel more secure about releasing viruses that attack insect pests, and if such viruses prove safe, this somehow bears on the widsom of feeding livestock on bacteria that secrete growth hormones. A substantial public relations effort is being mounted to create that impression. However, it is doubtful that there will ever be "enough data" on genetically engineered microbes, simply because each engineered organism is a separate case. And scientifically there is no obvious basis for requiring more data on an engineered virus than on a natural one. Valid concerns about the introduction of a particular organism, engineered or not, can be answered only by tests on the same organism or similar ones. The most useful tests are of course generic in that their results apply to groups of organisms rather than to single species, but the boundaries of the groups are set by criteria other than "engineered" versus "natural."

Proposals today that threaten to repeat the excesses of the guidelines in their most restrictive period are a cause of apprehension. Yet in 1989 a group of ecologists, stressing the complexity and poor predictability of interactions within biological communities, laid the groundwork for another restrictive regulatory policy.[13] Starting with the assumption that all introductions should be subject to regulatory scrutiny until a conscious decision is made to exempt them—a presumption of guilty until proved innocent—they developed multiple criteria that might form the basis for a hierarchy of risk categories. Such documents can be useful in identifying potential problems that might otherwise be overlooked, but they can also encourage a degree of regulatory overkill that may take much work to undo.

Scientific Responsibility

The response of molecular biologists from the early 1970s onward to the conjectural risks of recombinant DNA research has often been lauded as a model for how to handle new technologies and their possible attendant hazards.[14] The scientists involved have been praised for their honesty, their sense of responsibility, and their willingness to delay their own research while the issues were being aired. In this view later events have proceeded in orderly succession, with delays and inconveniences that may seem insignificant. A different perspective on the subject was given by Stanley Cohen:

> I think it is important to distinguish between perception of responsibility and actual responsibility . . . In retrospect, it seems to me that while the [Berg] letter was *perceived* as responsible, it was not really responsible at all. The most incriminating thing that any of us could have said at the time about recombinant DNA research was, not that there was any indication of hazard, not that there was even any valid scientific basis for anticipating a hazard, but simply that we could not say with certainty that there was not a hazard. The same thing could have been said about virtually any other kind of experimental endeavour. None of us who signed the "Berg" letter would have considered publishing a scientific paper without valid data, and yet simply on a basis of a lack of certainty that there was not a hazard, and in the absence of any real epidemiological input, a public statement was released.
>
> In doing so, I believe we showed poor scientific judgment, quite apart from any question of political judgment. However, because of the collective scientific credentials of the group, the recommendations of the letter were given credence out of proportion to their intrinsic merit.[15]

It might be more accurate to charge the authors not with poor scientific judgment but with a total lack of scientific judgment. They seem to have acted more as ordinary citizens responding to a new social or political issue than as experts addressing a technical problem.

Now what is really wrong with that? After all, aren't they citizens first and scientists second? If they sensed a possible threat to public safety, why shouldn't they speak out? The answer is very simple and very important. When a group of experts take a position on a subject within their area of expertise, both the scientific community and the general public have a right to expect that they have seriously evaluated the technical factors involved. This in no way implies that they should restrict themselves to purely technical advice or refrain from taking the broadest view possible. However, the world is full of well-meaning, intelligent people whose opinions deserve to be heard. A document about genetic engineering whose authors include several Nobel laureates in molecular biology is received differently from a letter on the same subject drafted by Jerry Falwell or Jane Fonda. It is proper to expect that the former group will give the subject the benefit of the one thing that only they or their colleagues can offer: considered scientific judgment.

It is appealing, and popular in some quarters, to admire the lofty goals that motivated the authors of the Berg letter and to attribute the monstrous network of restrictive regulations that confronted molecular biologists in the 1970s to the workings of the federal bureaucracy. Certainly both bureaucrats and political activists contributed to the problem. But the responsibility must rest primarily on the scientists who directed the Asilomar conference (and is shared by all of us who attended and voted on the recommendations that emerged); some of these scientists campaigned with great zeal for regulation of research. To be sure, the original Berg letter called for a "voluntary" moratorium; but this was accompanied by efforts to persuade institutions voluntarily to mandate compliance by individual investigators. Subsequently, at Asilomar the scope of regulatable experiments was extended to all work with recombinant DNA. Finally, the NIH, and later other federal agencies, agreed to deny funding to investigators who failed to comply, and eventually to institutions that employed even one such investigator, regardless of his or her source of research support. Both the members of the RAC and the NIH staff in charge of administering the guidelines conscientiously worked to execute their charges. The basic problem lay with the charge itself. In the end, the scientists who had initiated the action stood roughly in the position of the man who has his three wishes magically granted and belatedly comes to understand what he had asked for.

The NIH guidelines have remained nominally in force long after most RAC members, both scientists and nonscientists, have become convinced that they contribute nothing substantial to public safety, and far beyond the time when many RAC members would have voted to enact guidelines had they not already existed. The reasons are political rather than scientific: votes to keep research under the guidelines have been justified not by the expectation that any hazard will be avoided but by the assumption that "the public" is not yet ready to deregulate most of the research. Thus, the federal guidelines serve a preemptive function (politically, not legally), by reducing the pressure for regulation at other levels. If they do not protect society from harm, at least they help to protect Harvard from the Cambridge City Council and Genentech from the California State Assembly.[16]

Such posturing may be politically smart in the short run, but in the long run it may prove counterproductive. It conveys a false message as to the actual assessment of risk that knowledgeable people have. Genetic engineering is one of several areas in which politically aware scientists recognize that the general public—or sometimes members of Congress—perceive a problem where they themselves see little cause for concern. Starting from the position that an unfavorable public perception is itself a problem, many influential scientists respond by mounting an attack on the perceived problem, not on the misperception.

Conclusion

Veterans of the original discussions on the safety of recombinant DNA experience a sense of déjà vu in the current debate over planned introductions and may hope that a repetition of past mistakes can be avoided. Many scientists were disillusioned by the outcome of their own well-intentioned if ill-considered warnings of danger in the 1970s, and they are frustrated by delays currently imposed on planned introductions. This is especially so where the introductions appear completely harmless, and where the concerns expressed are unconnected to the fact that the organisms are engineered rather than natural.

However the legal issues may be resolved, there is strong scientific support for the position that no distinction between engineered and natural organisms is necessary or useful in evaluating potential hazards. Any large-scale dissemination of living organisms, natural or otherwise, may well entail at least some problems or dangers. The problems are commonplace and fall into categories for which regulation already exists, such as worker safety and environmental impact. The most important impact of genetic engineering may be to increase the numbers of organisms deliberately disseminated rather than to change the quality thereof. This may prove an appropriate time, therefore, to reassess regulations governing natural as well as engineered organisms.

Notes

1. For the legal action, see *New York Times*, 15 April 1981. Most other Serratia infections in the United States have also proved to involve strains different from that released in the Army tests (J. J. Farmer III, B. R. Davis, P. A. D. Gremont, and F. Gremont, *Lancet*, no. 2[1977]: 459–60). For a different perspective, see R. Y. Stanier, "The Journey, Not the Arrival, Matters," *Annu. Rev. Microbiol.* 34 (1980): 1–48.

2. S. E. Lindow, "Ecology of *Pseudomonas syringae* Relevant to the Field Use of Ice⁻ Deletion Mutants Constructed for Plant Frost Control," in *Engineered Organisms in the Environment: Scientific Issues*, ed. H. O. Halvorson, D. Pramer, and M. Rogul (Washington, D.C.: American Society for Microbiology, 1985), pp. 23–25.

3. There is no legal requirement for RAC approval, unless the release in question is performed by an institution that receives federal support. In practice, commercial firms have considered it advisable to comply voluntarily with the NIH guidelines.

4. A. Campbell, "Recombinant DNA: Past Lessons and Current Concerns," in *Introduction of Genetically Modified Organisms into the Environment: SCOPE 44*, ed. M. A. Mooney and G. Bernardi (Chichester: John Wiley, 1990), pp. 27–31.

5. See J. D. Watson, "Why the 'Berg' Letter Was Written," in *Recombinant DNA and Genetic Experimentation*, ed. J. Morgan and W. J. Whelan (Oxford: Pergamon Press, 1979), pp. 187–93. Documentation of the major landmarks is provided by J. D. Watson and J. Tooze in *The DNA Story: A Documentary History of Gene Cloning* (San Francisco: Freeman, 1981). See also W. Szybalski, "Much Ado about Recombinant DNA Regulations," in *Biomedical Scientists and Public Policy*, ed. H. H. Fudenberg and V. L. Melnick (New York: Plenum, 1978), pp. 97–140.

6. D. Stetten, Jr., "Valedictory by the Chairman of the NIH Recombinant DNA Molecule Program Advisory Committee," *Gene* 3 (1978): 265–68. Stetten captured the flavor of the Asilomar conference as I remember it better than any other observer: "I heard several colleagues declaim against sin, I heard others admit to having sinned, and there was a general feeling that we should all go forth and sin no more. The imagery which was presented was surely vivid, but the data were scanty. I recall one scientist presenting information on the difficulty of colonizing the intestinal tract with *Escherichia coli* K-12, but his presentation was given little attention. We were all, in effect, led down to the river to be baptized, and we all went willingly. I, for one, left the meeting enthralled. I had never been to a scientific meeting that so excited me."

7. See Morgan and Whelan, *Recombinant DNA*.

8. See B. D. Davis, "Bacterial Domestication: Underlying Assumptions," *Science* 235 (13 March 1987): 1329–35.

9. J. Rifkin, *Declaration of a Heretic* (Boston: Routledge and Kegan Paul, 1985), and Y. Baskin, "Genetically Engineered Microbes: The Nation Is Not Yet Ready," *Am. Scientist* 76 (1988): 338–40.

10. P. J. Regal, "The Ecology of Evolution: Implications of the Individualistic Paradigm," in *Engineered Organisms in the Environment*, pp. 11–19.

11. M. A. Martin, H. W. Chan, M. A. Israel, and W. P. Rowe, "Biological Activity of Polyoma-Plasmid and Polyoma-Phage Recombinants in Mice and Hamsters," in *Recombinant DNA*, pp. 195–204.

12. Quoted in J. L. Fox, "Emergent Scientific Principles for Deliberate Release," *ASM News* (Am. Soc. Microbiol.) 54 (1988): 355–59.

13. J. M. Tiedje, R. K. Colwell, Y. L. Grossman, R. E. Hodson, R. E. Lenski, R. N. Mack, and P. J. Regal, "The Planned Introduction of Genetically Engineered Organisms: Ecological Considerations and Recommendations," *Ecology* 70, no. 2 (1989): 298–315.

14. See, for example, the articles by Lewin, Weiner, and Thornton in *Recombinant DNA*.

15. In ibid., p. 296.

16. See also Baskin, "Genetically Engineered Microbes"; and Chapter 2 of Mooney and Bernardi, *Introduction of Genetically Modified Organisms*.

4

An Ecological Perspective

Simon A. Levin

Few issues in recent memory have caused such divisions in the scientific community as the deliberate release of genetically engineered organisms; yet examination of the underlying arguments reveals few differences among the scientific points put forward by ecologists, molecular biologists, and other scientists. It is true that a process of convergent evolution of viewpoints has occurred, developed through intense and sometimes acrimonious debate and discussion. Unfortunately, this process has left scars and distrust, and these remain as barriers to the full partnerships that should be developed. Position papers of various learned societies, such as the National Academy of Sciences (NAS)[1] and the Ecological Society of America (ESA),[2] have been prepared with the attention to detail usually reserved for legal documents, under the unfortunately correct assumption that each sentence may be scrutinized, quoted out of context, or misinterpreted. Such quibbling hampers progress, for it obscures the fact that basic agreement exists on the key issues. It is hoped that this volume will help us to go beyond this stage and encourage mutually beneficial partnerships.

Various recent documents seem remarkably uniform in their principal conclusions;[3] it is in the details and emphases that they differ.[4] All seem to agree that genetic engineering offers exciting possibilities for improved, less polluting approaches to environmental management, among other applications; that most introductions are likely to be benign, although a small number might present problems; and that the basis for regulation must be the product and its proposed use, instead of the process by which the product is created. Indeed, the language on these points is so similar that a sound basis exists for establishing a regulatory policy.

The potential for genetic engineering as a tool for environmental management is considerable. It is unfortunate that much of the debate over it has been represented as a struggle between those interested only in profit and those who would defend the environment against the profiteers. This view is both counterproductive and unfair to the majority of advocates of genetic engineering, whose goal is improvement of the human condition; but resolution of the problem has not been helped by an arrogance on the part of some genetic engineers, who seem surprised that anyone would question their objectivity in regulating themselves, and are content to malign rather than persuade their questioners. Similarly, the willingness—even eagerness—of some ecologists to participate in the scientific investigations necessary to resolve the issues has been portrayed by some as self-serving and a way to generate new research funds for ecology. This is an equally specious charge, which simply diverts attention from the substantive issues.

My own view is shared by the diverse group of scientists who attended the 1987 Bellagio meeting, sponsored by the Scientific Committee on Problems of the Environment (SCOPE) and the Committee on Genetic Experimentation (COGENE) of the International Council of Scientific Unions (ICSU) to examine the potential benefits and risks of deliberate release:[5] just as there are potential environmental costs that must be assessed in considering the deliberate release of an engineered organism, there may also be costs that attend a decision to delay the release. Thus, the use of genetically engineered alternatives to chemical fertilizers and pesticides could help to reduce the input of toxic chemicals into the environment; other genetically engineered organisms could degrade chemicals already existing in the environment. Furthermore, in the face of environmental change of anthropogenic and other origins, the techniques of genetic engineering may be used to increase the tolerances of agricultural and other desirable species to adverse factors, as well as to advance basic ecological and evolutionary studies. This point has been made repeatedly by a wide range of scientific groups, including the ESA. Thus, any delay in the implementation of these methods to address these problems bears its own environmental costs, which must be weighed against the possible costs and risks associated with introduction. To some extent, therefore, possible costs and benefits can be measured in the same currency. Of course, such simplification is not always possible; nor does this prescription for evaluation endorse carelessness in assessing potential costs, in the haste to bring desirable products to the market.

A familiar analogy to the case of engineered organisms is the development of new drugs, for which both potential entrepreneurial and human benefits must be weighed against costs and risks, known or unknown. Though the current system of drug regulation involves heartbreaking delays for those who might benefit from the new drugs and are prepared to take personal risks, it is designed to protect the populace from unexpected and possibly long-term side effects. In the case of environmental release of engineered organisms the potential benefits are less personal; the externalities are probably greater in that risks may involve others than the beneficiary. Yet the issues are similar in kind.

The intensity and emotional overtones of the debate have served to obscure the points of agreement, which are far more substantial than the points of disagreement. Where disagreements over deliberate release exist, they have more to do with how we should conduct our lives in the face of uncertainty than with what the scientific facts are. As Aaron Wildavsky argues elsewhere in this volume, the science matters in that it bounds uncertainty. That is, the critical issues of dispute are not scientific ones; they are personal convictions about the risks one is willing to take and the kind of world one prefers. In the case of deliberate releases of genetically engineered organisms, the risks are likely to be very small, certainly of the order that we have come to accept with the release of organisms modified by other means. That is consistent with the earlier point that the product rather than the process should be the basis for regulation.

Yet it is important to understand the perspective of ecologists, who are familiar with the difficulties of providing scientific advice in the face of uncertainty and where the costs and benefits of an activity are not distributed uniformly across all segments of society. In such cases it is wrong to oversimplify complex and unsettled issues by concealing them under the cloak of scientific certainty, or by confusing the public with some extremely complex computer model. Chastened by such experiences, ecologists view each new situation with skepticism, determined to ascertain whether what looks like certainty is what it seems. And with the process of debate and discussion that has attended the potential introductions of genetically engineered organisms has come a gradual relaxation of concern on the part of ecologists, who agree with many of the premises of those who believe that the great majority of introductions could cause no environmental problems. Yet the process for reaching these conclusions has been a very important one, allowing an independent confirmation of those premises. Despite the impatience of those who have had time to think about these problems a little longer, it is shortsighted to deny others the opportunity to consider the logical arguments carefully, to reach independent conclusions, and to disagree where necessary; only in this way can the broad scientific consensus be developed that is necessary to proceed responsibly with applications.

When scientific consensus can be reached that particular classes of introductions pose negligible risk, it should be possible to facilitate greatly the regulatory process. As the ESA states, "While we cannot now recommend the complete exemption of specific organisms or traits from regulatory oversight, we support and will continue to assist in the development of methods for scaling the level of oversight needed for individual cases according to objective, scientific criteria, with a goal of minimizing unnecessary regulatory burdens."[6] The ESA has presented a very preliminary set of criteria. Certainly, some of these criteria are debatable, and a final set could look quite different; but the process of developing consensus has begun.[7] I am much more sanguine than the ESA concerning the possibility of making broad exclusions, or at least of developing categories that require only minimal regulatory oversight; I suspect that some members of the ESA committee share this point of view. Thus, I do not interpret the above-quoted statement as meaning that it is not possible to agree quickly on such categories; rather, the members of the committee were not prepared to present a systematic classification of the necessary sort in a document developed for other purposes and on a short time scale. Attempts to go beyond the ESA statements have been initiated, however; the recent workshop at the Boyce Thompson Institute at Cornell University[8] makes such an attempt for genetically engineered plants; and the NAS has sponsored a broader effort to achieve this objective. We are capable of developing a classification system, and high priority should be given to the task. Such a system would allow us to concentrate regulatory resources on the most serious risks.

There is almost complete agreement that the great majority of introductions will be benign, but that some could be problematical. Generic arguments that all introductions present substantial risks are obviously fallacious; yet generic arguments for the safety of all introductions equally must be rejected. Elsewhere in this volume, Allan Campbell has pointed out that only a small fraction of all possible gene combinations can ever have occurred. Furthermore, even if a candidate gene combination has arisen before, the conditions under which it arose likely were different from present conditions. Problems have arisen with the introduction of non–genetically engineered organisms; examples include the disruptive influences of exotics and the heightened disease susceptibility of cereal grains resulting from the widespread use of new varieties. The introduction of genetically engineered organisms deserves the same attention as the introduction of organisms modified by other means. This is not to say that it will be possible to regulate all introduced organisms by the same statutes, convincing as the parallels may be. As Campbell argues, historical processes and other factors mold the nonuniform way in which society deals with comparable activities.

The fact that generic arguments on both sides cannot be supported does not, of course, pose a dilemma. Rather, it argues that nothing substantive can be said without dealing with specific cases. Plants, animals, and microbes differ greatly with respect to level and type of risk, and it is clear that similar distinctions are possible within each of these groups. The issue is not whether problem scenarios can be constructed; they can. The issue, rather, involves assessing the relevance of such scenarios to particular proposed releases.

The Seeds of Debate over Deliberate Release

The debate that has raged over genetic engineering for the past several years is complex in origin. In part it has arisen from scientific differences; however, in part it seems sociological in origin. The Asilomar conference of 1975 was a crucial event, perhaps the most crucial event, in shaping the views of molecular biologists. Those who were present felt that they participated in a major historical event, an event of the magnitude of Bishop Wilberforce's historic debate with T. H. Huxley at the Sheldonian Theatre, Oxford, concerning the evolutionary theory. Those who were not privileged to attend the Asilomar meeting, however, are to some extent disenfranchised: they have the option of endorsing the conclusions, not of Asilomar, but of those who attended and now know better. Otherwise, they are told, "We have been down this path before. You must learn from the mistakes of history." To an outsider to this fraternity, it seems that there is a mixture of pride at having taken steps that displayed high moral purpose, embarrassment that those steps now appear to have been overemotional and unscientific, and recognition that the conference and the Berg letter that preceded it led to restraints that are now deemed onerous and unnecessary. Clearly, it is argued, if scientists acknowledge that they do not know how dangerous their activities are, society can be expected to react even more cautiously. Many molecular biologists think that the Asilomar experience was an unfortunate one, and that today they are reaping the decaying fruits of a serious error.

The ultimate lesson of Asilomar is that things always seem clearer in retrospect than in prospect. The situation of deliberate release is not identical to that which existed in 1975, although a number of the key scientific issues remain the same. Those who now have the advantage of hindsight should use it to reduce the time needed for others to come to the correct conclusions, rather than take the view that the lessons of Asilomar make any further concern unnecessary.

When presented with the need to predict what will happen in some new situation, one naturally extrapolates from previous experience. This can and does lead to different conclusions for different groups of scientists, who will differ in the experiences that are most familiar to them, or that seem most relevant. There is a long history of safe introductions associated with agriculture, and this is offered as evidence that introductions produced through genetic engineering are also likely to be safe. The argument that the product rather than the process should be the focus of attention supports that view. On the other hand, ecologists point to problems that have been associated with both deliberate and accidental introductions of exotics. For the most part, these have differed in kind from the types of introductions currently being contemplated, but the relevance of this analogy is probably the most contentious scientific issue related to the matter of deliberate releases.

Be that as it may, it is clear that the relevance of the introduction of exotics to the assessment of risks associated with deliberate releases will vary from case to case, and in ways that are to a large extent predictable. The NAS committee thought that the introduction of exotics did not provide a good model when one was concerned with the genetic manipulation and reintroduction of species into environments in which they were already present, including agricultural systems. However, the model was obviously good if one were to introduce manipulated organisms into environments novel to them. This conclusion applies independently of the method used to modify the organism. Although the ESA argues that the distinction is not as easily made as the NAS committee believed, I believe that the conclusion is sound and provides an important first cut for purposes of classification.

Perhaps the core of the controversy has been a subtle distinction concerning the significance of the conclusion that the product, not the process, should be the issue. For the genetic engineer this means that there is nothing qualitatively new in genetic engineering, and that therefore no new vigilance is necessary. This view can be modified to recognize that as technology advances it might become easier to achieve certain objectives, or possible to do things that were not previously possible.[9] Still, as long as the product remains the focus, adequate protection seems equally available for genetically engineered organisms and for those modified by more conventional means. It is that tacit assumption that has encountered resistance. This inescapable corollary of the now accepted product-over-process reasoning makes many people squirm, because its implementation would make it more difficult to carry out introductions that until now have generally been regarded as safe. It might well mean, for example, that Gary Strobel would have met opposition to his proposed experiments on using *Pseudomonas syringae* to control Dutch elm disease, not because he introduced the modified Pseudomonas, but because he introduced the pathogenic Dutch elm disease organism.[10] This is not to suggest that the introduction of Dutch elm disease was risky under Strobel's particular conditions.

The fact that organisms modified by the new techniques do not differ fundamentally from those modified by conventional means is precisely what has caused ecologists to focus on the conventional concerns as the basis of their worries. This is, again, a corollary that one must accept once the product rather than the process has become the focus. The interpretation of "product" must include not just the organism but also the uses to which it will be put and the scale of its application. This argument has been made repeatedly and enjoys broad consensus. The recent position paper of the ESA supports the conclusions of the NAS committee. The précis provided by Mooney and Risser states that "ecological risk assessment of proposed introductions must consider the characteristics of the engineered trait, the parent organism, and the environment that will receive the introduced organism."[11] In the body of the document, it is stated that "we contend that transgenic organisms should be evaluated and regulated according to their biological properties (phenotypes), rather than according to the genetic techniques used to produce them."[12] This statement is virtually identical to those presented by the very different groups of scientists represented at Bellagio, or on the NAS committee.

Carrying the argument a step further, the ESA presents a list of specific concerns:

1. The creation of new pests . . .
2. Enhancement of the effects of existing pests through hybridization with related transgenic crop plants . . .
3. Harm to nontarget species . . .
4. Disruptive effects on biotic communities . . .
5. Adverse effects on ecosystem processes . . .
6. Incomplete degradation of hazardous chemicals, leading to the production of even more hazardous by-products . . .
7. Squandering of valuable biological resources [referring to the possibly accelerated evolution of resistance in insect pests through the use of *Bacillus thuringiensis* toxin][13]

Clearly, the level of concern that these various points should raise is debatable. The central point, however, is that these are not unique to genetic engineering per se. Each represents a point that is equally valid—however valid that may prove to be—for organisms modified by conventional technology. I agree that one is justified in raising these points once the product-over-process argument is accepted, as long as the same points are raised for other introductions. It is the product, including the patterns of use, that must be the focus of attention. Whether those patterns of use will be changed by the advances in technology remains to be seen, but it is not a problem as long as those patterns of use remain the focus of attention.

The ESA argues that "case by case review is currently the most scientifically sound regulatory approach."[14] *Case by case* is unpopular terminology because of the bureaucracy it suggests. It also should not be interpreted to exclude the development of classification schemes that will allow the individual investigator to take the initial steps in the case-by-case consideration.

A further problem with case-by-case review is that it deals with individual planned introductions rather than with patterns of use involving multiple introductions of a variety of organisms. Consideration only of the incremental effects of individual chemical pesticides is what has placed us in the current fix, in which the cumulative effects of multiple low-risk toxicants add up to substantial risks. Similarly, it is not just the individual introductions of organisms that deserve attention, but how society chooses to utilize this kind of increased capability to manipulate the environment.

Ecologists' Concerns

The concerns of ecologists regarding releases deserve serious attention. They can be organized, in analogy with the release of chemicals into the environment, into concern about the fate, transport, and effects of the genetic material introduced. There are major differences between the release of chemicals and the release of organisms: chemicals can become transformed or concentrated in the food chain; and organisms can reproduce. Yet though these differences flaw the analogy, they do not destroy it.

Unlike the objects of most ecotoxicological studies, the products of genetic engineering are unlikely to have direct adverse health effects on humans, unless, as in biological warfare, they have been designed specifically for that purpose. While there is a real potential for the use of biotechnology, including the new techniques of genetic engineering, by terrorists or belligerents to carry out biological warfare, it has little to do with the issues at hand. The possibility that such destructive organisms might be created by accident, as the result of engineering for some unrelated trait, was a real concern at the time of the Asilomar meeting. Today, however, those concerns have been laid to rest, and few scientists would regard this as a likely scenario. This consensus has been one of the positive results of the discussions and vigilance initiated at Asilomar.

Adverse effects on species other than humans are more likely, since some of the modified organisms will be designed to aid in pest control. Their undesirable effects on nontarget species present a familiar complication in any pest control program. However, such effects are less likely to result from biological agents than from chemicals, which in general are much less specific in their effects. On the other hand, introduced organisms, unlike chemicals, have the potential to reproduce, and to proliferate their genetic material in other ways.

Obviously, it is important to characterize the dispersal and geographical spread of organisms to be introduced into the environment. That is probably done easily for most genetically engineered organisms, provided the comparable information is already available for the unmodified species. If characteristics affecting transport have not been modified directly, then there is no reason to expect such changes to appear.

However, if a property such as herbicide resistance or frost tolerance is desired for particular species in particular environments but would not be desired over a broader taxonomic or geographical range, how can we be assured that biological populations will not proliferate the new genetic information beyond the target area? This concern has been exacerbated by the knowledge that the introduced material often will be found on conjugative plasmids, with substantial but poorly understood potential to spread among individuals and among species in natural environments.

It is generally agreed that favorable selection pressure is necessary for such spread; attention therefore should turn to evaluating whether such pressure exists. The historical ubiquity of plasmid exchange is another example of an issue that has been used to argue both points of view. For some, it is evidence that there is nothing new with transfers among distinct organisms; for others, it is used to argue that undesired exchanges are likely. That generic arguments can be used to draw such diametrically opposed conclusions is further proof of the need to deal with specific cases.

A separate, but related, issue is that we have the potential to spread a new product widely once it proves its value on a limited scale. This problem is a real one and has not been given sufficient attention. Once again, however, it is a problem that is not specific to genetic engineering. Each new instrument in our toolbox allows us to modify our environment even more, and the historical record is not testimony to our wisdom in exercising this power. The issue of scale of application, including international aspects, must therefore be an object for concern regarding any introduced organisms, independent of the means by which they have been produced. Of course, it must be recognized that the ease and efficiency of production will influence the potential for this form of proliferation. A case in point comes from the history of pesticides, as Harvey Brooks points out:

> When DDT was first introduced for agricultural purposes in about 1946, a great deal of careful technology assessment was done by entomologists, and it was concluded that the health and ecological hazards were minimal; but nobody at the time had any conception of the scale on which DDT would eventually be used, and the optimistic assessment resulted in relaxation of caution without considering the assumed limitation of patterns of use which underlay this assessment. The problem is that sufficiently attractive applications of a new technology may lead to loss of social control over its patterns of introduction and use.[15]

Introduced alien species have commanded the interest of ecologists for a half-century or more, and they provide evidence of the harmful effects that can occur when introduced populations increase without bound. It is true that most such introductions are unsuccessful, and deliberate introductions for biological control of insects and weeds often need to be repeated many times in order to increase the probability of success. This by itself does not put the matter to rest, since it is the low-probability but high-risk events that are the focus of attention. Others have discounted the introduced-species model, because most of the currently contemplated introductions involve species that are already found in the environments into which they will be introduced, but with a new trait added. However, there surely will be introductions of species that are not native, just as those introductions are made today with unmodified organisms. While recognizing the limitations of the introduced-species model, one is not justified in discarding it completely; it certainly will be relevant when the introductions contemplated are of non-native organisms.[16] Of course, once again this concern is not specific to organisms that have been engineered by new genetic techniques but applies to all such introductions, including those of unmodified organisms.

The issue of the fitness of engineered organisms is one in which I believe the available data do not support many of the conclusions being drawn. On the one hand, it is argued that engineered organisms are usually less fit than their wild-type sources, either because of some indirect cost associated with the new trait, or because engineering has not been precise enough to avoid deleterious effects at other loci. Both of these are effects that may be minimized as techniques become more efficient, and furthermore they cannot be counted on as protection in every case. In the overwhelming majority of cases I expect that the generalization will prove true; but it certainly is conceivable that some introductions—for example, those involving drought tolerance, salt tolerance, or herbicide resistance—will render the bearers more fit in particular environments. That indeed will be the intent of the engineering, in some cases. Thus, characterization of the fitness of the modified organism is essential to evaluating risk.

On the other hand, I do not find compelling the argument[17] that natural selection will tend to overcome the reduced fitness of introduced modified organisms, thereby making them more likely to spread undesirable traits. I agree that selection will tend to replace the introduced genotypes with more fit ones, if indeed the introduced ones were at a fitness disadvantage relative to the indigenous ones. But the most likely pathway by which this would occur is the elimination of the introduced trait, which is the culprit in reducing fitness, thereby returning the population to the wild types. This is indeed a problem, but only if the goal is to maintain the introduced type in the population.

Another concern of ecologists involves the transfer of genetic information to other populations, including those of different species, through mechanisms such as introgression (a special mechanism of gene transfer between plants) or the horizontal transfer of plasmids among bacteria. For plants, where the target population exists in the proximity of wild and weedy relatives, there is substantial potential for the transfer and spread of traits that enhance fitness, for example herbicide resistance. The consequence might be the need to increase the use of other herbicides in order to control the newly created pests. This, however, applies equally when herbicide resistance is introduced into a target population by more conventional methods. Moreover, it is likely to be a problem only for a few taxonomic groups; in the United States most row crops are not surrounded by wild and weedy relatives, sorghum being an exception.[18] For forage grasses in the United States, and for row crops as well as forage grasses in South America, there is the potential for introgression, and introductions must be approached more cautiously.

Plasmids have attracted considerable attention because many engineered traits will be plasmid-borne, and because it is clear that historically there has been a great deal of exchange of plasmids through conjugation, transduction, or transformation among distantly related bacterial species. This phenomenon adds great complexity to bacterial systematics. The spread of antibiotic resistance because of strong selective pressures is the best-known and most startling consequence. Like viruses or pathogenic bacteria, plasmids can in theory be maintained in populations by horizontal transfer even if the characteristics they bear are slightly detrimental to their hosts. That possibility, however, is conjectural, and its relevance remains a fascinating topic for research; in general, engineering of a plasmid gene in a way that does not increase the fitness of the bearer is very unlikely to result in proliferation of that trait. Certainly, if the engineering does increase the fitness of the bacterial cell in a particular environment, as occurs "naturally" for plasmids conferring bacterial resistance in the presence of antibiotics or heavy-metal ions, horizontal transfer has the potential to accelerate greatly the spread of the introduced genetic information.

The issue of the displacement of indigenous microbial species by introduced ones seems not to be a major concern, largely because of the functional redundancy of microbial species relative to their ecological roles. This is not to say that displacements will not occur, but simply that their ecosystem consequences are not likely to cause harm. Given the state of our knowledge of microbial taxonomy, and the fact that only a small fraction of microbial species can be identified, little more can be said at this point. As our knowledge of these species increases, many of the generalizations being made now will have to be modified. In any case, the need is clear for an increased attention to the ecology and systematics of those microbial species that play fundamental roles in driving and regulating critical biogeochemical processes.

The final point that I want to address concerns the importance of the scale and frequency of introductions, where my view may differ from that of the editor of this volume. A selectively neutral mutant has a probability of fixation that is proportional to its initial frequency. That fact, which is central to the neutral theory of evolution, follows easily from elementary mathematical arguments: under selective neutrality every gene has equal probability of being the sole ancestor, eventually, of all members of the population (at some distant time). A slight selective advantage will alter this result quantitatively, but not qualitatively; the initial frequency remains important in determining the probability of fixation. For small populations (and localized introductions) fixation is not likely to occur if there is substantial input of genes from other populations. Larger-scale introductions involve populations that are more closed, and for these the above arguments are appropriate. This line of argument assumes no nonlinearities, which may operate to introduce thresholds of a different nature. In any case, the conclusion is that the scale of application is important to determining whether a particular introduction is likely to spread; the world is not deterministic, and a slight mean selective advantage is neither necessary nor sufficient

for that genotype to spread. A single recombinant bacterium is very unlikely to be the progenitor of a large population, even if it has a slight
selective advantge; a planned local introduction is therefore more likely
to be controllable than a widespread commercial application.

This conclusion has significance for the problem of containment,
which is manageable for small introductions but assumes different dimensions for large-scale applications. Thus, I am in agreement with the
views of the ESA, the NAS, and Campbell in supporting small-scale and
controlled field tests as a necessary step in getting the products of biotechnology into the field, but in urging caution where large-scale commercial applications are contemplated.

Conclusion

Virtually every responsible body that has considered the issue of
risks associated with deliberate releases of genetically engineered organisms has drawn the same conclusion. Simply stated, it is the product
(and how it is used) and not the process that provides the proper basis for
evaluating the associated risks. This means that there are risks, but
they are the same in kind as risks that we have learned to deal with as
the result of centuries of experience with conventionally modified organisms. Those risks must be evaluated in association with proposed
introductions, but we have sufficient knowledge to develop categories
that can facilitate risk assessment. The criteria that can form the basis
of such a classification system include properties like pathogenicity, potential for horizontal transfer and spread or introgression, dispersal
characteristics, and whether the introduced organism is native to the
target environment; and preliminary steps are being taken toward development of more detailed protocols.[19]

Some have suggested that one might distinguish among introductions according to whether the organism has novel genetic features that
could not have arisen through natural events; I believe that such a criterion would be unsound. On the one hand, using this as a basis for decision making would not provide sufficient protection, in that it would
not exclude destructive but naturally occurring organisms, such as the
AIDS virus. Furthermore, many proposed modifications may have occurred naturally, but so rarely or under such conditions that they could
not become established. The criterion is also flawed by its focus on process rather than product.

Logically, there is no need or basis for separate regulations based on
the method by which organisms have been modified, although as Campbell notes, the crazy quilt that has created our regulatory patchwork
may make regulatory uniformity impractical. The objective should be

to develop an appropriate scheme for evaluating risks associated with introductions, with the emphasis on the properties of the engineered organism in relation to the environment into which it will be introduced. We have the capacity to develop a procedure that will safeguard the environment without unduly hampering the development of products that pose negligible risk. And we must do so expeditiously, for there are substantial environmental costs associated with delaying the development of nonpolluting alternatives to chemical fertilizers and pesticides. Ecologists, genetic engineers, applied biologists, and the public at large have common interest in facilitating the application of biotechnology to solving environmental problems.

Notes

It is a pleasure to acknowledge the critical reading and comments by Harvey Brooks, Allan Campbell, Bernard Davis, Arthur Kelman, and Richard Mack. This chapter is ERC-207 of the Ecosystems Research Center at Cornell University, and was supported in part by the U.S. Environmental Protection Agency Cooperative Agreement CR812685 with Cornell University. The work and conclusions published herein represent the views of the author, and do not necessarily represent the opinions, policies, or recommendations of the funding agencies.

1. National Academy of Sciences, *Introduction of Recombinant DNA–Engineered Organisms into the Environment: Key Issues* (Washington, D.C.: National Academy Press, 1987).

2. J. M. Tiedje, R. K. Colwell, Y. L. Grossman, R. E. Hodson, R. E. Lenski, R. N. Mack, and P. J. Regal, "The Planned Introduction of Genetically Engineered Organisms: Ecological Considerations and Recommendations," *Ecology* 70, no. 2 (1989): 298–315.

3. Ibid.; NAS, *Introduction of Recombinant DNA–Engineered Organisms;* Boyce Thompson Institute for Plant Research, *Regulatory Considerations: Genetically Engineered Plants. Summary of a Workshop Held at Boyce Thompson Institute for Plant Research at Cornell University, October 19–21, 1987* (San Francisco: Center for Science Information, 1988); Office of Technology Assessment, *New Developments in Biotechnology (3). Field Testing Engineered Organisms: Genetic and Ecological Issues* (Washington, D.C.: OTA, 1988); S. A. Levin, "Safety Standards for the Environmental Release of Genetically Engineered Organisms," special combined issue of *Trends Ecol.* 3, no. 4 (1988), and *Trends Biotechnol.* 6, no. 4 (1988): S47–S49; H. A. Mooney and G. Bernardi, eds., *Introduction of Genetically Modified Organisms into the Environment: SCOPE 44* (Chichester: John Wiley, 1990).

4. D. T. Kingsbury, "Deliberate Release III: It's Not What You Say, It's the Way That You Say It," *Trends Biotechnol.* 7, no. 5 (1989): 110.

5. Mooney and Bernardi, *Introduction of Genetically Modified Organisms.*

6. Tiedje et al., "Planned Introduction," p. 298.

7. A. M. Campbell, "Deliberate Release: Will the Ecologists' Report Disturb the Biotechnology Ecosystem?" *Trends Biotechnol.* 7, no. 5 (1989): 109–10.

8. Boyce Thompson Institute, *Regulatory Considerations.*

9. Tiedje et al., "Planned Introduction," p. 300.

10. Levin, "Safety Standards," p. S48.

11. H. A. Mooney and P. G. Risser, "The Release of Genetically Engineered Organisms: A Perspective from the Ecological Society of America," *Ecology* 70, no. 2 (1989): 297.

12. Tiedje et al., "Planned Introduction," p. 302.

13. Ibid., p. 301.

14. Ibid., pp. 307, 310.

15. H. Brooks, personal communication, 1989.

16. NAS, *Introduction of Recombinant DNA–Engineered Organisms,* p. 14.

17. Tiedje et al., "Planned Introduction," p. 303.

18. R. N. Mack, "Invading Plants: Their Potential Contribution to Population Biology," in *Plant Populations: Essays in Honour of John L. Harper,* ed. J. White (London: Academic Press, 1985), pp. 127–42.

19. NAS, *Introduction of Recombinant DNA–Engineered Organisms;* Tiedje et al., "Planned Introduction"; Boyce Thompson Institute, *Regulatory Considerations.*

5

An Environmentalist Perspective

Margaret Mellon

Environmentalists were active in the first wave of debate about laboratory research with recombinant organisms, and their interest has increased throughout the subsequent years.[1] Now at least four national environmental organizations are involved with the issue, and local groups are becoming active in a number of states.[2]

The main reason for the current wave of public interest in genetic engineering is the prospect of commercial production of a broad range of genetically engineered organisms. Commercialization entails the large-scale introduction of engineered organisms into the environment, which poses broader risks than those associated with laboratory research. In addition, issues other than risk are emerging. Genetic engineering is no longer simply a research tool. Advanced techniques are poised to become the basis of new industries with the potential to transform major sectors of human society and the environment. Among other things, genetic engineering offers the increased domestication of forests and fisheries, the replacement of food plants and animals with cell cultures, and expanded, nonagricultural uses of plants and animals. Each of these developments represents a choice, not an inevitability. Environmentalists have joined the debate to work to control risks and to bring an environmental perspective to these choices.

Engineered organisms are being offered as solutions to problems high on the environmental agenda, like cleanup of pesticides and toxic wastes. One might therefore expect environmentalists to be as anxious as industry to get this technology under way. Part of the reason they are not lies in the history and the values of the environmental movement, and part in its experience with previous "miracle" technologies.

Environmental groups, running the gamut of conservation and environmental quality issues, place a high value on preserving human interactions with nature in an increasingly urbanized and mechanized society. Deeply conservative, they prefer the organisms of the earth as 4 billion years of evolution have given them to us. Currently, those organisms are under enormous pressure from pollution, development, and habitat loss. Against that background, the prospect of a new technology with new power to manipulate and further subdue nature does not have much appeal.

At the same time, environmentalists are practical. They understand the need to exploit nature in order to sustain human life and the positive role genetic improvement has played in developing plants, animals, and microbes for food, fiber, and drugs. They do not doubt that the new techniques of biotechnology could provide important benefits in the future.

Even so, they do not advocate a rush to embrace the technology. Experience with earlier "miracle" technologies counsels otherwise. Experience with synthetic chemicals, for example, has demonstrated that benefits usually come at a price. For the benefits of the seventy thousand chemicals now in commercial production, that price has turned out to include polluted air and water, pesticide-contaminated ground water, manufacturing and transportation accidents, ten thousand abandoned hazardous waste sites, an eroding ozone layer, and the possibility of rapid global climate change.

The miscalculations of the past are constantly before the environmental community. Daily contact with such issues as chlorofluorocarbons, pesticides, and air toxics brings home the fact that technologies are double-edged swords. Synthetic chlorofluorocarbons, for example, which became popular for a wide variety of industrial and commercial uses because of their extreme stability and low toxicity, are the major cause of the thinning of the stratospheric ozone layer, and they also contribute to global warming. This realization has not led environmentalists to reject technology out of hand. But they do not assume that technologies come without risks. And they are concerned that a couple of decades from now, when the biotechnology industry is producing tens of thousands of organisms for commercial use, its risks might approach those of chemicals.[3]

Perhaps most important, environmentalists do not assume that solutions to the world's problems depend entirely on new technologies. In fact, they often see promising nontechnological approaches being overlooked in the rush to the next technological bandwagon.

The New Technology

Biotechnology refers to a variety of activities. It is sometimes used to denote any practical use of living organisms, from water buffalo plowing rice paddies to industrial enzyme production. Sometimes it is used less broadly to denote activities involving heritable modifications of organisms by traditional and by modern genetic methods. For many environmentalists, the subset of greatest interest is further limited to organisms modified by powerful modern genetic techniques and intended for outdoor release.

Thus, the intensity of environmental interest varies across the spectrum of activities generally considered biotechnology. Currently, the major focus of interest is commercial-scale planned introductions into the environment. Large-scale introduction increases the likelihood that engineered organisms will find suitable habitats and become established in the environment, and hence the likelihood that mistakes will be irremediable. Other applications of the technology, such as monoclonal antibodies, protein engineering, and chemical fermentation processes, are of lesser interest because they either do not involve living organisms or else use organisms not likely to survive outside of controlled conditions.[4]

At the core of both the promise and the risks of genetic engineering is the power to produce combinations of genes not found in nature. Modern gene transfer techniques allow scientists to bypass biological mechanisms and directly transfer functional genetic material to host organisms.[5] For example, human genes can be transferred to cows, or plant genes to bacteria. These techniques vastly increase the ability to generate organisms with new properties. Using them, scientists have already produced sheep that secrete human proteins in their milk for use as drug-production facilities; plants with genes that sequester metals for use with fertilizer made of metal-contaminated sludge; and tobacco plants with the genes of fireflies that make them glow for use as test systems for light-based organismal markers. What interests people most about genetic engineering, however, is not its present but its future. If functional genes can be taken from any organism and transferred into any other organism, the potential reach of the technology is very broad.

At present, numerous impediments reduce the range of practical gene transfers to far less than the theoretical limit. For example, only a tiny fraction of the genes in nature have been identified and isolated; existing transfer techniques limit potential hosts; the techniques are restricted to small blocks of DNA that usually contain only one gene of interest; the final locations of transferred genes on host chromosomes are random, so insertions of new sequences may interfere with the function of existing genes; and organisms have functional limits to their tolerance of the disruptive effects of added genes. But most of these restrictions are technical in nature and will not impede the technology for long. The library of genes available for transfer grows almost daily. At the same time, the list of possible host organisms is increasing as scientists develop new techniques for gene transfer, such as the recent technique to propel genes into cells on tiny metal projectiles.[6] Finally, as the understanding of gene function and regulation continues to improve, progress can be anticipated in both the number of genes per transfer and the availability of mechanisms to avoid disruptive effects.

Thus, we have begun on a long journey of technical feats involving biological organisms. We are not very far down the road. But the breakthrough has been made. We are headed toward a future where scientists can purposefully and efficiently create genetic novelty by transferring and rearranging genes and groups of genes. Because this new ability involves the rearrangement of components, it resembles engineering more than breeding. Indeed, *genetic engineering* is the term that has caught on to describe the implementation of modern genetic techniques.

There has been much discussion about whether genetically engineered organisms differ in significant ways from naturally occurring or traditionally bred organisms. While they have many similarities, they are different in three important ways. First, engineered organisms can contain combinations of genes that are highly unlikely to arise by natural mechanisms of gene exchange. Second, compared with natural processes or traditional breeding, the category of engineered organisms is open-ended, in theory leading to organisms with properties and uses that are beyond our imagination. And third, engineered organisms are produced by direct, often mechanical, gene transfer procedures, many of which are wholly unrelated to natural processes. These differences are a cause for concern because our relationship to other organisms is affected by whether, and how, we have interfered with their lives and reproduction. The concerns raised by genetic engineering fall into three main categories: imaginary risks, genuine risks, and broader concerns.

Imaginary Risks

An issue that lurks in the background of the genetic engineering debate is the possibility of producing monsters. The term *monsters* usually refers to organisms that are more than simply dangerous. Neither the lion nor the cholera organism is a monster, although both can kill people. The term usually refers to plants or animals that are grotesque or unnatural in appearance and also possess harmful properties—killer tomatoes or wasps the size of dachshunds.

Perhaps because of a belief that any response may give undue credence to the concern about monsters, government regulators and proponents of biotechnology tend to ignore the issue altogether. This may be unwise, since the idea has a considerable hold on the public imagination. In fact, current genetic technology does not pose a real risk of creating monsters. On the contrary, as long as transfers are confined to one or two genes, most products of genetic engineering are going to be utterly familiar. A tomato might inadvertently be made to overproduce a toxic substance, but it will still be a tomato.

If the technology should progress to the point of multigene transfers capable of producing unfamiliar organisms, the risk of producing organisms that would be considered monsters would increase but would still be very small. Even multigene transfers would still produce primarily familiar organisms. Because there would be no commercial incentive to make a monster, it would have to be the result of some sort of mistake. Although not impossible, the likelihood of such a mistake producing an organism that is simultaneously unfamiliar, dangerous, and able to survive is very small.

Genuine Risks

But disposing of monsters does not dispose of the concerns about genetic engineering. There are genuine risks associated with the production and introduction of genetically novel organisms. These risks are not generic; that is, there are no properties or risks attaching to an organism just because it has been engineered. Instead, risks arise from properties of the added genes, the effects of new combinations of genes, and specific environmental situations. These risks can be both direct and indirect.

Direct risks fall into two categories. One includes new properties conferred upon the host by added genes, which could pose risks to human or animal health. Examples are crops into which insect-toxin genes have been added as pesticides. The U.S. Department of Agriculture has received several applications to field-test toxin-containing tomato plants.

The other category of direct risk includes ecological risks, which encompass the effects of genetically novel organisms through interaction with other organisms. The Ecological Society of America (ESA) has reported a number of potential effects of released organisms to be avoided. Among these are the creation of new pests, damage to non-target species, disruptive effects on existing biotic communities, and adverse effects on ecosystem processes.[7] A report from the National Academy of Sciences (NAS) published in 1987 also noted the ecological risks of releases of engineered organisms, such as the displacement of existing species of fish.[8] Because ecological effects emerge from complex interactions of organisms, they are often unexpected and difficult to predict. Even in retrospect, for example, it is difficult to explain why kudzu and gypsy moths have been so successful, while so many other introductions of exotic plants have failed. Potential ecological effects also vary widely, depending on whether the organism is a plant, an animal, or a microorganism, how it has been engineered, and where and in what quantities it will be released.

One example of the potential ecological effects of engineered organisms is engineered fish. Scientists in laboratories around the world are currently engineering fish to contain foreign genes.[9] Scientists have succeeded in transferring two foreign genes into different carp—one gene coding for a growth hormone, the other for an "antifreeze" protein intended to enable the fish to survive in cold waters.[10] If introduced, intentionally or accidentally, into the aquatic environment, either of these engineered fish could displace or disturb native populations of fish. The antifreeze gene, for example, could enable a warm-water fish to expand its range into colder waters. Or the rapidly growing engineered fish might displace normal fish by depleting the supplies of food organisms in a pond before the normal fish were large enough to catch them.

There is little reason to suppose that existing species are so well adapted that newcomers would necessarily be unable to displace them. In fact, introduced non-native fish have often driven native species to the brink of extinction.[11] Since engineered fish are like non-native, or exotic, fish in that they possess traits that are novel in relation to the environment, they appear to pose the same serious risks. In view of such concerns, a workshop of the American Fisheries Society recommended that "all types of genetically engineered fish should be classified as exotic forms when developing regulations for their use in either natural or aquacultural settings."[12]

Fish engineered by new techniques may also pose other kinds of risks. For example, the cancer-causing retroviruses being used as vectors for moving genes into fish could pose risks of disease. At this point, prior to the introduction of a single fish, it is premature to rule out the possibility of new risks peculiar to the engineered fish. But it is not necessary to postulate so-called unique risks to justify concerns about engineered fish. The nonunique risks of displacement, disruption, and loss of wild stocks are more than enough to warrant concern.[13]

Besides these direct environmental effects, the use of genetically engineered organisms may have indirect environmental consequences. The best example is herbicide-tolerant plants. The development of major crops that are tolerant to chemical herbicides seems likely to lead to the increased or prolonged use of these dangerous agricultural chemicals.

A final category of genetic engineering risks is unknown risks. Scientists are modifying a broad spectrum of self-replicating entities in new ways. It is too early in the game to foreclose the possibility of consequences to these activities that have not yet been realized and perhaps cannot yet be conceived.

Broad Concerns

The horizon of environmental interest in engineered organisms is not limited to product-specific risks. The advent of biotechnology also raises broad environmental concerns in a number of areas. One area is the social and economic impact of the technology, especially in agriculture. In the United States and around the world, biotechnology products will affect the number of farmers, the choice of agricultural practices, and the control of agriculture. Since agriculture represents a major land use in almost all societies, its structure and practices have major environmental ramifications. In the United States, a major concern is that genetic engineering will accelerate the trend toward the environmentally destructive practices of industrial agriculture, such as high chemical inputs that pollute land and ground water.[14] In the third world, the displacement of local export crops, such as vanilla and coffee, by cell-culture substitutes engineered elsewhere would cause major social and agronomic damage. For example, if French genetic engineers succeed in developing cultures to produce gum arabic, they could cause the demise of the Sudanese gum-tapping industry, which is already under pressure. Until recently, gum arabic ranked second only to cotton in generating foreign currency for Sudan.[15]

Another broad area of environmental concern is the potential use of genetic engineering to produce pollution-tolerant organisms. An example is the development of genetically engineered fish that could grow in the acidified lakes of the Adirondacks. Such fish would probably find a ready market. Beyond the risks that the modified fish would pose of escaping and disturbing ecosystems, their use would raise other questions. The availability of fish adapted to polluted lakes might undercut the determination to clean up air and water. Their comparatively low cost might tempt some people to stock lakes with acid-tolerant fish instead of controlling emissions of pollutants. However tempting in the short run, adapting biota to polluted environments is the path toward life in a permanently polluted environment.

The availability of genetically engineered organisms could have complex implications for wildlife, which is another matter of wide environmental concern. On the one hand, higher-yielding crops could conceivably lead to a more efficient agriculture, allowing land now under cultivation to revert to wildlife habitat. On the other, new crops adapted to what are now marginal agricultural lands might encroach even further on habitats suitable for wildlife.

The issue of our relationship with nature, especially with wildlife, is also raised by genetic engineering. Although the issue is speculative now, eventually the technology will make it possible to engineer wild animals like wolves, eagles, or bears. Genetic engineering might be suggested, for example, to make wild animals easier to protect or to manage. Since we value such animals precisely for their independence and ability to survive without us, our relationship with nature would be poorer if many of the wild organisms had been genetically tailored to suit human purposes.

An environmental problem of overriding concern is the ultimate direction of the technology. An important characteristic of genetic engineering is its open-endedness with respect to the variety and novelty of its products. Few imagined cows as pharmaceutical factories, a development already upon us; no one can imagine the products of the genetic engineering industry of 2025. By contrast, the possible products of both nature and classical breeding are constrained by mechanisms of natural reproduction. By sticking with these methods, we are assured that change will come somewhat slowly and that the final result will not depart greatly from either the familiar or the natural. By casting aside the barrier of natural mechanisms, however, we opt for a world where the look, behavior, and uses of organisms will have few bounds, and where we shall have few guideposts.

Much of the nervousness about the technology derives from this sense of starting down a path that may lead in unwanted directions. Some of the uneasiness can be attributed to fear of the unknown; some to the understanding that many techniques will be readily transferable to humans; and some to our adverse experience with other technologies. Whatever its source, concern about the future colors the current debate about applications of the technology.

The Need for Regulation

Few people are comfortable with the notion that scientists should be able to construct engineered organisms of any kind and introduce them anywhere in any quantity. There is broad agreement that the technology should be regulated for its environmental risks. The major issues are how extensive the regulatory oversight should be and whether new legislation should supplement or replace existing laws.

Because of the diversity of products and activities expected from the technology and the large number of potentially applicable statutes, the regulatory issue is a complicated one. The federal statutes applicable to engineered organisms are inadequate to protect against the risks of environmental introductions.[16] The major statutes pressed into service as the so-called federal framework—the Federal Plant Pest Act, the Federal Insecticide, Fungicide, and Rodenticide Act, and the Toxic Substances Control Act—are flawed in their jurisdiction, data-gathering requirements, and enforcement provisions. The flaws and gaps are so serious that the patchwork approach should be scrapped and new, comprehensive legislation enacted.

The cautious position on regulation would be to subject all introductions of genetically engineered organisms to federal oversight. The purpose of the regulation would be to screen out the minority of organisms that might present problems. Eventually certain categories of organisms would be expected to qualify for exemption from review.

So far, however, although the concept of risk-based categories is widely endorsed, no scientific body has succeeded in devising such categories.[17] The ESA suggested an approach based on the characteristics of the parent, the nature of the genetic modification, and the attributes of the environment into which the organism would be released. But the ESA recommended against establishing categories right now because of the diversity of products that can be developed and the complexity of predicting their ecological fate.[18] The NAS developed a framework for decision making similar to the ESA approach, but also provided no specific categories.[19] For the time being, the current system of case-by-case review is the scientifically sound approach. Although relatively rare, problems could be large and virtually irremediable. As the gypsy moth and kudzu illustrate, once organisms are free in the environment they are almost impossible to eradicate. It is simply too early to exempt categories of engineered organisms from oversight and control. Workable schemes for categorizing organisms according to risk will emerge from accumulated experience and from research stimulated by risk assessments.[20]

That deliberate introductions of engineered organisms raise distinctive environmental concerns is illustrated by a product now under review, the biopesticide being developed by Crop Genetics International, Inc. (CGI). Using modern genetic techniques, CGI has moved the gene for the insect toxin produced by *Bacillus thuringiensis* into an endophyte, a bacterium that grows inside plants. The company hopes that the pesticide-producing bacteria infused into corn and rice seeds will provide plants with protection against major insect pests. This is a worthy goal, but there are two kinds of risks.

First, there are possible risks to human and animal health. Although certain naturally produced *Bacillus thuringiensis* toxins have been demonstrated to be extraordinarily safe for mammals, there is no proof that the engineered toxin is the same as the naturally produced toxins that have been tested. Without proof of identity, the products of the engineered organism cannot be presumed to have the same safety characteristics as the tested molecules. There are scores of these naturally produced *Bacillus thuringiensis* toxin molecules, which vary in host range and toxicity.[21] Even where the amino acid sequence is the same in the engineered as in the natural toxin, differences in added sugar moieties could affect protein function.[22]

Second, there are potential ecological risks. Since the parent bacterium was not even discovered until the early 1980s, the first task of the review process was to require basic studies on the natural history of the organism. So far, the early tests have revealed that the parent organism, in addition to growing in corn and rice, can also grow in a wide variety of wild plants. Furthermore, it can be transmitted from crops to other susceptible plants by mechanical means, as might occur during harvesting.

A readily transmissible, toxin-containing organism might pose a number of risks. If the endophyte containing the toxin gene is transmitted to and survives in nontarget plants, and if the toxin gene persists in the bacterium, the endophyte could confer resistance to some insects on infected weeds. This advantage could cause the toxin-containing endophytes to be selected for and become established in wild plants. Where insects are an important control on weed populations, insect-resistant weeds might become more difficult to eradicate. Also, where rates of toxin production are high, infection with the bacteria might render some plants lethal for nonpest insects like butterflies.

On a different level, commercial use of the product on large-acreage crops like corn and rice might provide sufficient exposure to the toxin to elicit resistance in pests.[23] If resistance to the toxin does develop, it would mean the loss of one of the safest pesticides. This concern is exacerbated by the fact that many other companies are engineering the same toxins into field crops, vegetables, and trees. These issues require government scrutiny before the organism is released into the environment in commercial quantities for application to millions of acres of farmland year after year.

Many people oppose regulation on the grounds that it will impede the international competitive position of the United States. This argument assumes both that regulation necessarily impedes commercial success and that other countries will not regulate genetic engineering as stringently as the United States. Neither is the case.

Regulated industries are often immensely profitable. In fact, the early U.S. successes in biotechnology have occurred in precisely those areas—drugs and diagnostics—that are heavily regulated. The notion that other countries are not restrained by concerns about engineered organisms is also not true. The major competitors of the United States are more, not less, conservative in their approach to the technology. A recent survey of readers of a popular Japanese science magazine—most of whom were educators, government officials, and students—revealed pervasive apprehension about genetic engineering.[24] In line with that attitude, Japan has barred the introduction of all genetically engineered organisms into the environment, pending development of a regulatory scheme.[25] In Europe, the British Royal Commission on Environmental Pollution recently recommended new legislation to establish a system under which *all* releases of genetically engineered organisms would require a permit.[26]

Design of an oversight scheme covering the diversity of potential products expected from biotechnology is a formidable challenge. Ideally, the system would require only one review by one agency for each introduction. It would work toward a set of risk-based categories to use in scaling the intensity of review, and local committees would be used to oversee low-risk introductions. As experiments went forward, the program would compile a database on the results of early introductions, to refine the review categories and improve the subsequent risk assessments.

The chances of achieving the ideal system would be far greater if the government were starting from scratch to design a system for biological organisms rather than trying to adapt statutes already enacted to regulate chemicals. Whether approached by enacting new legislation or reworking existing statutes, however, regulation cannot be avoided. At bottom, the public will neither accept nor support a technology that presents uncontrolled risks. Credible, efficient regulation would advance, not retard, the biotechnology industry.

Uncertain Benefits

Much has been made of the tendency of opponents of genetic engineering to inflate the risks of release. And there have been cases in which risks were overblown. Yet much the same can be said about the claims for benefits.

Proponents of the industry paint a rosy picture of the benefits of genetic engineering. Foremost among the claims are that the technology will wean agriculture from chemicals and feed the world's burgeoning population. Chemical-dependent agriculture and world hunger are real and pressing problems, but genetic engineering cannot be counted on for solutions.

For example, look at the first two major products of the agricultural biotechnology industry from the standpoint of promised benefits rather than risks. The two are bovine growth hormone (BGH) and herbicide-tolerant crops. Both of these products are disappointing to anyone expecting substantial environmental benefits from agricultural biotechnology.

Bovine growth hormone, also called bovine somatotropin (or somatotrophin), is considered a product of biotechnology because it is produced by bacteria engineered to carry the BGH gene. Such bacteria, grown in large vats, are the only way to produce large, easily purified quantities of BGH. BGH is a substance that, if injected daily, stimulates cows to give more milk on less food. But the product is not likely to help feed the world. To begin with, the primary market for the product is not the third world, but the first. Because the products are available only at first-world prices, the existing dairy surpluses in the United States and in Europe have done little to relieve hunger in the third world. It is unlikely that the greater surpluses resulting from BGH will have greater impact.

Even in first-world countries the benefits, if any, will be limited. In the United States, for example, consumers will not benefit, because prices at the grocery reflect government support systems more than supply and demand. Taxpayers will not benefit, because taxpayers pay billions of dollars every year in subsidies to maintain an already overly successful dairy industry. To address the milk glut, the government in 1986 and 1987 paid most of the bill for a $1.8 billion program to induce 14,000 farmers to slaughter 1.5 million cows and calves.[27] Least likely of all to benefit are the small dairy farmers, who will be driven off their farms as fewer cows suffice to produce the same amount of milk. Concerns about the impact of BGH on the dairy industry have prompted Vermont, Minnesota, and Wisconsin to consider legislation either banning or labeling BGH milk. The only obvious benefits of BGH are to the chemical companies, which will sell the chemical to those farmers who survive the contraction in the industry. American agriculture has long assumed that productivity is synonymous with progress. As the case of BGH demonstrates, that assumption needs a critical reassessment.

The second major product of the agricultural biotechnology industry, herbicide-tolerant crops, are being developed by large, transnational chemical companies. Worldwide, at least twenty-eight enterprises have launched more than sixty-five research programs to develop herbicide-tolerant crops. Fifteen major crops are involved, including cotton, corn, soybeans, wheat, and potatoes.[28] The engineered crops being made resistant to the company's own herbicides will surely increase the market for the chemicals. Industry analysts expect herbicide-resistant plants to prove to be big business, mainly for pesticide and herbicide producers.[29]

Herbicide-resistant products run directly counter to the promise of a reduced dependence on agricultural chemicals. Rather than weaning agriculture away from chemicals, these products will shackle it to herbicide use for the foreseeable future. Herbicides represent an estimated 65 percent of the chemical pesticides used in agriculture.[30] By failing to replace herbicides, the biotechnology industry greatly restricts its potential for reducing overall chemical use.

While these two products offer no hope of substantial environmental benefits from agricultural technology, there are alternatives to engineered organisms that do promise substantial reduction in overall pesticide use. These alternatives fall under the rubric of sustainable agriculture. Sustainable agriculture employs a variety of agricultural practices—such as crop rotation, intercropping, and tilling fields to produce long ridges of soil on which plants are grown—that control pests without application of synthetic pesticides and fertilizers. Applied by knowledgeable farm managers, these techniques work to make farms both profitable and environmentally sound.[31] Growing different crops in successive growing seasons, for example, dramatically reduces pests by sequentially removing the hosts on which the pests depend. With fewer weeds or insects to contend with, the farmers save the increasingly expensive cost of purchased chemical pesticides. Once thought impractical, sustainable agriculture is rapidly gaining support in national policy forums, including the NAS.[32]

The environmental advantage of such an approach is clear. Crop rotations that reduce the pests reduce the need for pesticides—chemical or biological—now and into the future. Thus, developing sustainable practices should be the highest priority for agriculture as it moves toward the twenty-first century. Although a minor part of the picture, biotechnology could help by developing new crop varieties usable in sustainable systems. For example, faster-germinating, cold-tolerant crop varieties could enhance low-input, sustainable systems by enabling more effective weed control.[33] Such engineered products would fulfill the promise of biotechnology and would benefit farmers, the environment, and the rural economy alike.

The Role of the Public

The public has a vital role to play in the development and direction of biotechnology. Public education about the technology is needed to enable people to evaluate its risks and to participate in decisions about its use. The public education needed is not simply a campaign to calm fears and render the technology acceptable to the public. Doubts about the technology do not come only from unfounded fears born of too many late-night horror shows. Unreasonable fears exist and should be countered, but there are numerous genuine risks and nonrisk concerns connected with the advent of the technology.

An educated public can evaluate the risks and weigh them against the benefits. Just as important, the public should have a say in choosing among the benefits of the technology. In agriculture, for example, the public should have the choice between products and research that support sustainable agriculture and those that support high-chemical systems.

In short, there is no need to oppose the technology across the board. It must be evaluated application by application. Generally speaking, with proper controls the technology should prove useful in basic scientific research and in commercial fermentation systems to produce drugs and specialty chemicals. For example, use of engineered bacteria to produce substances, like insulin, that are otherwise unavailable in needed quantities is to be welcomed.

As for the environmental applications of the technology, they should not be endorsed until more is known about both risks and benefits. Environmental releases of engineered organisms, particularly on a commercial scale, pose potentially significant risks. Thus, caution is needed as well as strict government oversight of planned introductions.

In agriculture, the most promising and exciting developments for the future are in sustainable agriculture rather than high-technology engineering. Biotechnology products could play a positive role in the transition to an environmentally sound agriculture, but only if product development is subordinated to the principles of low chemical inputs and sustainability.

Guarantees are needed of public involvement in selecting among particular applications of the technology. Not all applications will prove beneficial, and society does not have to take the good with the bad. It should be prepared to reject those applications that have unacceptable risks or undesirable socioeconomic impacts or that threaten a preferred relationship with nature. Society need not regard technology as a sort of train that cannot be stopped once it gets on track. We can and must learn to evaluate technologies and, where appropriate, to just say no.

A well-developed public research effort is needed to complement private research and development efforts. Publicly funded programs, freed from commercial constraints, can investigate innovative approaches to problems, such as crop rotation to control pests, that depend on skilled management rather than product inputs. Publicly funded research can also develop products, such as narrow-spectrum biopesticides, that do not offer a sufficient return to justify private investment for research and development. Such programs also provide opportunites for public involvement in setting the direction of technological advance.

Finally, it would be unwise to brush aside concerns about genetic engineering in headlong pursuit of its "tremendous" benefits. Cautious development of the technology will give us the opportunity to avoid its risks and ensure that it serves—not dictates—our social and environmental goals.

Notes

1. Groups like the Friends of the Earth and the Natural Resources Defense Council were involved in the early debate about laboratory research. The Environmental Policy Institute began working on agricultural and seed patent issues as early as 1980.

2. The four groups are the National Wildlife Federation, the Environmental Defense Fund, the Environmental Policy Institute/Friends of the Earth/Oceanic Society, and the National Audubon Society. Examples of local groups are the Minnesota Food Association, the Wisconsin Family Farm Fund, and the California Biotechnology Action Council.

3. The chemical industry is in fact similar to the biotechnology industry in important ways. Not only do the corporate players in the two industries overlap, but just as chemical technology advances exponentially, success with one engineered organism opens doors to many others. Just as the chemical industry took off after World War II and within four decades was producing over seventy thousand commercial chemicals, the biotechnology industry can be expected within a few decades to produce tens of thousands of products.

4. This does not mean that industries involving these processes raise no environmental concerns. The Soviet Union, for example, has recently abandoned its entire single-cell protein program (involving over a hundred facilities) in the wake of incidents in which air emissions of highly allergenic proteins caused thousands of cases of bronchial asthma in surrounding communities. A. Rimmington, "Biotechnology Falls Foul of the Environment in USSR," Bio/Technology 7 (1989): 783. Other concerns over fermentation facilities involve waste disposal and long-term occupational hazards.

5. M. Mellon, Biotechnology and the Environment: A Primer on the Environmental Implications of Genetic Engineering (Washington, D.C.: National Wildlife Federation, 1989), pp. 17–23.

6. C. Stanford et al., "Delivery of Substances into Cells and Tissues Using Bombardment Process," *Particulate Sci. Technol.* 5 (1987): 27–37.

7. J. Tiedje, R. K. Colwell, Y. L. Grossman, R. E. Hodson, R. E. Lenski, R. N. Mack, and P. J. Regal, "The Planned Introduction of Genetically Engineered Organisms: Ecological Considerations and Recommendations," *Ecology* 70, no. 2 (1989): 298, 301.

8. National Academy of Sciences, *Introduction of Recombinant DNA– Engineered Organisms into the Environment: Key Issues* (Washington, D.C.: National Academy Press, 1987), p. 19.

9. N. Maclean et al., "Introduction of Novel Genes into Fish," *Bio/ Technology* 5 (March 1987): 257–61.

10. R. Dunham et al., *Experiments to Evaluate Common Carp Containing the RSV Rainbow Trout Growth Hormone Gene (rtGHg) for Inheritance, Expression, and Biological Effects of rtGHg in Experimental Ponds: Request for Review:* An Application to the Agricultural Biotechnology Research Advisory Committee, Submitted to USDA (February 1989).

11. NAS, *Introduction of Recombinant DNA–Engineered Organisms,* p. 19.

12. *Fisheries* 13, no. 6 (1989): 8.

13. *Fisheries* 11 (March/April 1986).

14. C. Hassebrook and G. Hegyes, *Choices for the Heartland: Alternative Directions in Biotechnology and the Implications for Family Farming, Rural Communities, and the Environment* (Walthill, Nebr.: Center for Rural Affairs, 1989).

15. A. Hollander, "Biotechnologies Are Tricky to Manage in the Third World," *Conservation Foundation Letter* 6 (1988): 1, 2.

16. Mellon, *Biotechnology and the Environment,* pp. 37–49.

17. NAS, *Introduction of Recombinant DNA–Engineered Organisms,* p. 23; and Tiedje et al., "Planned Introduction," p. 310.

18. Tiedje et al., "Planned Introduction," pp. 307–11.

19. National Research Council, *Field Testing Genetically Modified Organisms: Framework for Decisions* (Washington, D.C.: National Academy Press, 1989), pp. 65–76, 123–31.

20. Tiedje et al., "Planned Introduction," pp. 307, 310.

21. J. Morrison, "Soil Yields 72 New Varieties of a Natural Pest Control," *Agricultural Research,* January 1988, p. 14.

22. Scientists at a seminar on BT reported the surprising finding that purified BT toxin molecules were glycosylated and that the degree of glycosylation appeared to be related to function. J. Fox, "Natural Pesticide a Challenge to Manipulate," *Bio/Technology* 7 (1989): 1004.

23. F. Gould, "Evolutionary Biology and Genetically Engineered Crops," *BioScience* 38 (1988): 26–33.

24. D. McCormick, "Not as Easy as It Looked," *Bio/Technology* 7 (1989): 629.

25. "Japan Fights Fears of Gene Manipulation," *New Scientist* 118, no. 1610 (1988): 26.

26.   Royal Commission on Environmental Pollution, *The Release of Genetically Engineered Organisms to the Environment* (London: Her Majesty's Stationery Office, 1989), pp. 57–64.

27.   K. Schneider, "Biotechnology's Cash Cow," *New York Times Magazine*, 12 June 1989.

28.   Rural Advancement Fund International Communiqué (Pittsboro, N.C.: November 1987).

29.   M. Ratner, "Corp Biotech '89: Research Efforts Are Market Driven," *Bio/Technology* 7 (1989): 337, 341.

30.   Environmental Protection Agency, *Pesticide Industry Sales and Usage: 1986 Market Estimates* (Washington, D.C.: EPA, 1987).

31.   D. Granatstein, *Reshaping the Bottom Line: On-Farm Strategies for a Sustainable Agriculture* (Stillwater, Minn.: Land Stewardship Project, 1988).

32.   National Research Council, *Alternative Agriculture* (Washington, D.C.: National Academy Press, 1989).

33.   Hassebrook and Hegyes, *Choices for the Heartland*, pp. 16, 23.

6

Public Policy

Aaron Wildavsky

In the debate over regulation of biotechnology, few come out for the policy in which they believe or believe in the policy that they support. They all tell the truth as they understand it. But in order to appear "moderate" or "reasonable" and thereby gain public support, the rival sides avoid following the strict implications of their arguments. Thus, those who believe in the monstrous potential of biotechnology—in disasters so bad that no good from the same technology could possibly overcome them—should be arguing not for stronger regulation but rather for prohibition of biotechnological research and its application. And those who believe that biotechnology leads to results no different from those spontaneously occurring in nature or practiced for centuries by breeders and hybridizers, so that the wondrous good it will do is bound to exceed vastly the damage it causes, should not even accept existing regulation but should rather argue for its abolition.

The cause of the profound disagreements about regulation of biotechnology might seem to lie in the uncertainty surrounding this new technology and its application. That the disputants differ about the consequences of various applications of biotechnology is true. This disagreement, however, has deflected attention from their general agreement that the chances of inadvertent harm are remote, even exceedingly remote. Where the sides differ is on the degree of harm that might be done in such an unlikely but still possible event.

This difference in the awfulness of remote events is not due to uncertainty, for such a degree of uncertainty is inherent in life.[1] It would be akin to saying that we cannot agree on night following day because of Humean skepticism that what has always happened before will necessarily happen again. The difference is rather due, in Michiel Schwarz and Michael Thompson's locution, to "contradictory certainties," namely rival views of how the world works.[2]

The disagreements among those who seek to influence public policy on genetic engineering derive from the protagonists' cultural construction of nature. People's preferences for different ways of life influence their perception of how the world works and therefore the promise and pitfalls of a particular form of biotechnology. Thus, the moral justifications necessary to maintain different ways of life are related to the protective ideas that their adherents develop about how nature works. The differing perceptions of the effects of biotechnology make sense in terms of the shared values and desired social relations of the people who hold them.

Moreover, adoption of either looser or stricter forms of regulation, ranging from total prohibition to no regulation whatsoever, is likely to have different consequences for human health and safety. The crucial question is whether the burden of proof of the planned introduction of genetically engineered organisms should be placed on the proponents, who must show a particular application to be safe before it can be tried, or on the opponents, who must demonstrate harm before they can stop whatever is proposed.

Both the perceptions and the consequences of technologies deserve study. Without the former we cannot understand political behavior, or why people want what they want and consequently act as they do. Without the latter we are less likely to appreciate the substantive results of the chosen policies.

A clue to the contested character of biotechnology comes from a decision by the Massachusetts Institute of Technology to close down its small Department of Biotechnology on the perfectly sensible grounds that "biotech is a term, not a discipline." Indeed, so numerous and varied are the disciplines in some way related to biotechnology, the university argued, that a department composed of a mere twenty professors would find it impossible to encompass them.[3] The very definition of *biotechnology* is a matter of controversy, because definitions emphasizing its similarity to techniques that have existed since time immemorial, like breeding and cross-breeding animals and plants, suggest that nothing very new (and therefore very worrisome) is going on; on the other hand, definitions stressing genetic manipulation conjure up in willing minds images of Mary Shelley's Dr. Frankenstein creating a new (and therefore frightening and possibly uncontrollable) monster-like creature.

Biotechnologies do not involve the creation of life. Genetic engineering refers to the ability to manipulate to some degree the signals called genes that regulate the development of life forms, and particularly to take a gene from one organism and insert it into another, so that it becomes part of the effective gene system of the recipient. There are three different categories of genetic engineering: whole organism, cellular, and molecular.[4] Whole-organism plants and animals are repeatedly crossed and selected to obtain more desirable traits, such as endurance or resistance to pests. Manipulation of whole organisms has limitations. It is slow and expensive, although such technologies as the transfer of embryos, hybridization, and artificial insemination speed things up. Breeding is also largely limited to the gene pool within a species. And along with whatever desirable trait emerges may well come unwanted, indeed unknown, traits. Only trial and error will tell whether a hybrid corn that otherwise does so well will also be susceptible to epidemics.

Cellular genetic engineering begins by growing entire plants from cultured cells. Cells from different parents are then fused to create a hybrid that possesses the characteristics of both parents. Thus, cellular fusion permits genetic transfers across as well as within species. The possibilities of control with cells, though greater than with whole organisms, are not yet precise.

That honor goes to molecular genetic engineering, which uses recombinant DNA techniques to transfer specific genes within and across species. What is more, subject to some limitations, genes may be mutated, rearranged, or produced synthetically and then inserted into the desired host.[5] It is precisely this new power that has given rise both to such great hope for improving the human condition and to such great concern that if something goes wrong when these organisms enter the environment, the repercussions could be disastrous.[6]

Rival Cultural Views

The debate over biotechnology resembles the debate over other areas of risk. There is general agreement, for example, that AIDS and ionizing radiation can kill. Disputes occur along the edges of uncertainty. With AIDS, this uncertainty involves the question of casual transmission and contact tracing; with radiation, it involves the effects of cumulative low doses that are not quite known. The same penumbra of uncertainty affects the release of engineered organisms into the environment. Everyone agrees that nothing has happened . . . yet. The probability of damage is also conceded to be low. But here, at the edges of uncertainty, disagreement occurs over the low probability of potentially highly damaging consequences.

The science does matter. It has already removed a good deal of uncertainty, and there is no question of the good that might still be done by genetically engineered organisms. Because the disagreement begins where science ends, however, the extent to which uncertainty is perceived and the mechanisms by which it is resolved are central to the policy debate over biotechnology.

On the surface, the debate over introduction of genetically engineered organisms appears strange, for the potential benefits are immense, the dangers remote and so far unrealized. A clue to why disagreement over policies grows amid substantial agreement on the facts lies in the appropriation of this issue within a much larger conflict, the debate over the desirability of industrial technology for human health and environmental safety. Four different theories have been advanced to explain the rise of opposition to industrial technology.

The first theory bases such opposition on knowledge. Presumably people fear a technology because they have good reasons, based on existing knowledge, for doing so. If this is true, then it should follow that the more knowledge individuals have, the more fearful they are. But available evidence suggests otherwise. The most expert professionals, such as nuclear engineers and microbiologists, generally regard the relevant industrial technologies as acceptable.

The second hypothesis bases opposition to industrial technology on personality. Some people presumably are risktakers, while others are risk-averse. Even casual inspection, however, reveals that people who evidence risktaking in one area, say nuclear power or the buildup of radon in tight, energy-efficient houses, are risk-averse in regard to other kinds of dangers, such as AIDS or living near a nuclear plant. The personality theory fails to explain why dangers alleged to stem from industrial technology are singled out for regulation, while other dangers are not.

The third hypothesis relates positions on industrial technology to economic self-interest. This works in the sense that most of those who engage in biogenetic research are opposed to stringent regulation, while most ecologists, whose skill would be in greater demand under more stringent regulation, favor it. But self-interest is too narrow an explanation, excluding most other citizens and activists involved in the debate.

The failure of these three simple explanations makes it worthwhile to try a fourth, complex explanation, relating attitudes on industrial technology to culture. Within very wide limits, people choose what to fear in order to support their preferred way of life (or culture).[7] This hypothesis may appear startling to those who believe that the nature of things forces itself on our consciousness. But to those like myself who adhere to a social constructionist view, in which individuals are the active organizers of their consciousness, imposing their own meaning on phenomena, this cultural thesis is normal. There are four cultures that select their perceptions so as to sustain themselves: the hierarchic, individualistic, fatalistic, and egalitarian.

Adherents of each of these cultures create through interaction a mental model of physical nature, of how the world works. Since hierarchies are full of rules and regulations with many gradations, the adherents of hierarchy, or hierarchists, conceive of human nature as so bad that it can be improved only through the efforts of good institutions and their expert administrators. By contrast, egalitarians, who wish to diminish differences among people and who therefore find social stratification immoral, see human nature as basically good unless corrupted by evil institutions. To allow some people to suffer even for the purported benefit of a larger number of others is to place unwarranted trust in existing inegalitarian institutions. For individualists, people are neither good nor bad but self-regarding. To behave better, they require institutions that embody appropriate incentives. And for fatalists, who conceive of physical nature as random and, hence, intervention as pointless, human nature is similarly unpredictable.

Just as people seek to make their ideas of human nature fit their vision of proper social relations, they also try to make their notions of physical nature fit that vision. Of course, they cannot make nature do whatever they want. But they can manipulate their conception of nature so that it supports their preferred way of life. As David Bloor puts it,

> Men use their ideas about Nature and Divinity to legitimate their institutions. It is put around that deviation is unnatural, displeasing to the gods, unhealthy, expensive, and time consuming. These instinctive ruses map nature onto society. Nature becomes a code for talking about society, a language in which justifications and challenges can be expressed. It is a medium of social interaction . . . Classificatory anomalies . . . take on a moral significance. By these hidden routes they acquire the connotations of irregular social behaviour, which makes a response to them all the more urgent. One response is to "taboo" the anomaly which violates the classification, declaring it an abomination and seeing it as a symbol of threat and disorder.[8]

There is no better way of putting down ideas and the way of life behind them than by saying that they are unnatural, monstrous, a threat to all life. And there is no better way of countering this challenge than to say that what one proposes, and hence the way of life one favors, is natural, pure, unadulterated, coming straight from Mother Nature herself.[9] Formulas are right at hand: capricious nature versus cornucopian nature versus fragile nature versus perverse or tolerant nature.

The sets of convictions people have of how the world is and of how best to act in that world justify the pattern of social relations that is integral to their way of life.[10] Each model of nature, in consequence, soon recommends itself as self-evident truth to the people whose way of life renders them partial to that particular representation of reality (see Table 6.1). Though it is useful to classify these models according to cultural type in regard to technology, an individual would not necessarily adhere to the same culture and its model of nature in other social contexts and in regard to different kinds of problems.

Cornucopian nature knows few limits to ingenuity. Since unfettered individuals can always create more of what is wanted, resources can confidently be expected to grow. This model therefore justifies bold experimentation. It upholds a life of competitive individualism. Fragile nature is the opposite. It presumes a terrifyingly unforgiving world, in which the least jolt may trigger a catastrophic collapse. Strict regulation, perhaps prohibition, is required to prevent Mother Nature from wreaking vengeance on us for using our physical powers to exploit her, just as we use our inequalities to harm those who have less power or money. This model posits an egalitarian way of life. The egalitarian/fragilist would argue that initial inequalities tend to be amplified in the absence of collective restraints, while the individualist/cornucopian would argue that a rising tide floats all ships. Perverse or tolerant nature not only assumes the existence of jolts in moderation but also endorses the need for experts to avoid knocking nature out of its accustomed

Table 6.1 Models of nature and the corresponding cultures

Cornucopian Nature	Fragile Nature
The Individualist: People underestimate the power of human ingenuity to overcome apparent limits.	*The Egalitarian:* Tread the earth lightly or it will crush humanity, just as the powerful always crush the weak if they have the chance.
Perverse or Tolerant Nature	*Capricious Nature*
The Hierarchist: Expert knowledge is required to prevent the system from being destabilized by human intervention.	*The Fatalist:* Life is a lottery, there is little the individual can do about it, and what you do is as likely to hurt as to help.

place. Such a model justifies a hierarchical way of life, in which authority inheres in officially approved positions. With these three models, learning is possible, though each model disposes its proponents to learn different things. In the wasteland of capricious nature, however, the world is essentially random, full of surprises both pleasant and unpleasant. It is futile to try to anticipate and provide contingency plans for all of them; the better course is to roll with the punches. A culture of fatalism, in which nothing individuals do can positively affect their future, makes the most sense when nature is random.

Insight into why one culture cannot accept the models of another culture may be had by asking what would happen if they did. If egalitarians adopted the model of nature as cornucopian, they could no longer justify sharing because there would always be more for everyone. If individualists portrayed nature as fragile, they could not justify ceaseless trial and error. If hierarchists believed that nature was capricious, they could no longer legitimate expertise. And if fatalists believed in any of the other models of nature, they would be unable to rationalize their passivity and would be bound to do something about problems.

Cornucopian nature is the model that readily recommends itself to the entrepreneur, that advocate of the free market. Perverse or tolerant nature is the bureaucrat's model. It says that to follow the rules and the experts ensures that all will be well, but to act above one's station will bring catastrophe. Capricious nature is the model that reconciles to their fate those who find themselves squeezed out by all these institutional forms. And fragile nature is the myth of the egalitarian group. Believing that society ought to be characterized by equality, they adopt a view of nature that makes inequality unnatural, something imposed by force through evil institutions. Egalitarians therefore blame individualists and hierarchists for disturbing the natural order among human beings. That inegalitarian cultures disrupt the harmony of nature is just another proof of their immoral character.

Science is a form of competition, the competition over ideas. It is strongly individualist. But ideas also have to be retained, and this requires a modicum of hierarchy, though not so much hierarchy as to stop rival ideas. The egalitarians, as the cultural critics of existing authority, also aid innovation by challenging received wisdom. Science emerged out of the challenge to hierarchy that began in the early Middle Ages. It therefore depends on the same alliance of individualists and egalitarians that spawned capitalism. When, as is happening today, this underlying consensus is shattered—when egalitarians begin to believe that science is allied with forces of oppression generating inequality—the legitimacy of science through its technological products comes under attack.

What is amazing is not that the debate over deliberate introductions of engineered organisms reflects these perspectives, but that the antagonists behave as if they had stepped out of these categories. Often it appears as if the categories themselves do the talking. The protagonists act out perfectly the aspects of the models of nature.

The Natural versus the Unnatural

Virtually the entire debate between proponents of "full speed ahead" with biotechnology and those who wish to delay it can be described in two critical terms: *natural* versus *unnatural*. Competition over which culture is superior is carried on in significant part over claims as to what is natural, or in accord with the laws of society and nature, and what is unnatural, or monstrous, repulsive, inhuman. When adherents of a particular way of life convince others that it is the only natural way, so that its perceptions of how the world works become the way the world actually is, then they have won.

The clue to unlocking the codes telling us which claims about the world justify what kind of culture comes from the fit between the model of nature and the way of life. The trick is to follow the cultural meaning through the scientific discourse. Thus, the view that planned introductions of genetically engineered organisms are "natural" belongs to the competitive individualists who find nature to be cornucopian. They exude competency: they know the science, know the practice, know that things will turn out well, and know what to do if and when things go wrong. Those who perceive planned introductions as potentially monstrous support the model of fragile nature. Their imaginations run toward scenarios of how things can go wrong. Just as the nature-is-cornucopian individualists picture how the good consequences of release can make up for the bad, the nature-is-fragile egalitarians are adept at envisaging how monstrous events can take place.

These cultural distinctions do not mean that all or most biotechnologists are individualistic and that all or most ecologists are egalitarian. They mean that if the political views of these two groups could be ascertained, ecologists would be far more in favor of equality of condition among people or between animals and people, and biotechnologists would be far more supportive of individualistic inegalitarian outcomes in power, status, and income.

First, consider the claim of proponents of the natural. In a protest against the notion of producing a "monster," or perceiving a thing as so far outside the normal that taboos must be erected against it, the microbiologist Bernard Davis contends that all the steps in the recombinant DNA technology occur as processes in nature; investigators have simply increased their efficiency.[11] The ability to alter living things, Gerald Campbell continues, is not dangerous because unprecedented, for "mankind has been altering living species since before recorded history began." There is no need to worry about Ice⁻, for instance, "since bacteria lacking the gene that produces ice crystal formation have always been present in the environment, created there by natural mutations."[12] The biologist Winston Brill stresses that microorganisms in the environment "have naturally exchanged genes . . . A tremendous amount of gene transfer occurs naturally . . . even between kingdoms." In fact, "what scientists create through genetic engineering is minuscule and ecologically insignificant compared to what occurs continually and randomly in nature."[13]

If genetic transfer were novel, it might be hazardous. So the National Academy of Sciences (NAS) is careful to say that "genetic exchanges brought about by unconventional nonsexual means occur often in nature." They are "like a breeder's new variety of a flower." Rejecting the frequent analogy between recombinant DNA and the introduction of "alien organisms" that wreak harm, like the gypsy moth and the starling, the NAS contends that the modified organism resembles its parent (in other words, is not anomalous), except that it is "less able to survive."[14]

The modified organism's inability to survive, Davis argues, is predicated on evolutionary theory, which predicts that recombinants from distant organisms are less likely to be dangerous than are those from closely related organisms. The reason is that survival is compromised when new genes have not co-evolved with old ones. Thus, distant gene insertions "can be expected to produce noncompetitive (or even nonviable) monstrosities, rather than dangerous monsters." For the same reasons, Davis finds fault with the analogy to rabbits in Australia or the kudzu vine in America. These organisms overran their new environment "because it lacked the balancing forces present in the first . . . whereas the altered organism lacks that advantage."[15] A more reassuring analogy is to domestic species whose alterations increase their usefulness but not their ability to do better in the wild than the original strain. In sum, Davis theorizes, "the greater the genetic manipulation the more dependent rather than the more dangerous the organism will be."[16]

Evidently, how one conceives of the balance of forces in nature, called *homeostasis*, helps to determine whether one finds the alteration of genes more or less worrisome. If Davis is correct in asserting that "evolution does not select for maximization of some powerful trait, but rather for a *balanced, coadapted* genome, in which the parts fit each other well, just as in any effective machine," then moving around a gene here or there should pose little danger.[17]

The partisans of biotechnology do not argue that all ecological niches are full or that no harm has come from traditional techniques. They do claim that while engineered organisms may cause some troubles, these will be normal in that they are unlikely to go "beyond those we accept and manage in traditional practices."[18] Indeed, these partisans have thrown down the gauntlet to their opponents by claiming that the intentional creation of, say, a new weed would be a formidable task. As the NAS explains,

> The traits contributing to pathogenicity include the ability to attach to specific host cells, to resist a wide range of host defense systems, to form toxic chemicals that kill cells, to produce enzymes that degrade cell components, to disseminate readily and invade new hosts, and to survive under adverse environmental conditions outside the host. Together with the need to compete effectively with many other microorganisms for survival, these traits form an impressive array of requirements for pathogenicity.[19]

Thus, according to Waclaw Szybalski, creating an altered organism running amok is as likely as inadvertently creating "a television set by a random mixing of electronic components."[20]

Similarly, Davis maintains that since the number of bacteria in a shovelful of soil could well exceed the human population of the earth, and since these bacteria exchange genes, albeit slowly, the chances of being done in naturally are much greater than being done in by a few experiments. It is not easy, in other words, for a bacterium to become a bad bacterium if it is born a good one.[21] And if something untoward should occur, there is every likelihood of an early warning.[22] Biohazards have not in fact happened.[23] Campbell draws the moral: "One animal cannot take over without being made to do so by humans."[24]

Having disposed to their satisfaction of "stumbling into a catastrophe without warning," the advocates of a fast track for biotechnology point to its potentially immense benefits.[25] Their most telling argument is that the major crops of American agriculture, from rice to soybeans to wheat, are not native to North America, while today's corn is far from the original variety. Moreover, the largest part of American livestock, both beef and poultry, has foreign origins and has been subject to intensive manipulation. These benefits are a product of trial and error, and they could not have been obtained if their developers had first had to prove that they would cause no harm. In short, say the proponents of biotechnology, there is no need to worry, and public policy should not be based on "fear of imaginary monsters."[26]

The proponents of more stringent regulation agree that the likelihood of harmful consequences from the introduction of a genetically engineered organism into the environment is a product of its ability to survive, multiply, disperse, compete, and impact others. The probability of any one of these factors being detrimental, both sides agree, is low, and the probability of all of them coinciding is exceedingly low. The problem lies elsewhere.

The problem, according to ecologists, is that low risk is not zero risk. Events of low probability but high danger cannot be ruled out. Therefore, efforts should be undertaken to reduce these unlikely but possibly devastating hazards.[27] Martin Alexander, for example, argues that while modified organisms may be at a disadvantage,

> some of the newly acquired traits may aid the novel organism in coping with abiotic stresses and with such biotic stresses as competition, biological toxins, parasitism, and predation. Thus, if genetic engineering can confer on an organism a trait that makes it a better competitor or more resistant to toxins, predators, or parasites, the organism might not suffer appreciably even if it had to expend more energy to synthesize additional DNA or enzymes.[28]

To quote one of Alexander's aphorisms, "No tank never leaks," meaning that things can always go wrong.

There are always deviant cases. A gene from a crop could be transferred naturally to a related weed. In fact, Steve Olson warns, "this seems to have occurred several times in the past, as in the case of weedy Johnson grass and sorghum."[29]

An indication of how things could get out of control with higher organisms comes from the ecologist Frances Sharples: "Additions of nonindigenous organisms can influence the structure (population size and species diversity) and function (energy and material dynamics) of ecological communities through a variety of mechanisms that sometimes displace or destroy indigenous species." The ecological harm that "domesticated species" can do is illustrated by goats or rabbits going wild and doing "massive damage." Furthermore, "the premise that all possible gene combinations have already been tested in nature cannot be true." There are too many possible combinations. Indeed, "nature is full of harmful phenomena."[30]

According to Robert Colwell and other ecologists, "The idea that nature has a tight homeostatic balance is not borne out by field observation or experimentation." Small genetic alterations have indeed caused large harmful consequences. At times, genetic engineering may even "be *more* likely to produce new or more troublesome weeds."[31]

The proponents of additional regulatory precautions do not believe that most planned introductions would have a seriously harmful impact. They do believe that since there are few precedents for this sort of change, it is difficult to predict its effects. But if natural forces are indeed so elusive that they can be detected only through intensive field testing, and if it is too dangerous to release genetically altered organisms for such tests, we can never find out whether these hypothetical hazards will actually manifest themselves.

Because the debate over planned introduction of engineered organisms is carried on by rival teams trading cultural charges—"We are natural, you are unnatural; You are unnatural, we are natural"—it may seem that the science is pure rationalization. Not so. Each of these claims makes sense within the understanding of those who make it, even though both claims cannot be true. Rational science does matter, but because the science has become intertwined with deep differences over ways of life, the rival claims are going to be more than usually difficult to resolve.

The Double Bind

In the war of analogies, microbiologists and ecologists have tended to take different sides. For ecologists, the dominant analogy is nature the practical joker, which is subject to Murphy's Law (anything that can go wrong will). For microbiologists the dominant analogy is to lawlike nature, emphasizing the regularities of the natural world. Ecologists tend to cite hard cases, while biologists speak of principles. Thus, Alexander observes, "it is commonly believed, especially by microbiologists whose specialties are not related to the environmental sciences, that microorganisms do not persist in environments in which they are not native."[32] Davis responds, "It is remarkable that we can still be arguing, on the basis of analogies rather than firm scientific principles or evidence, about hypothetical disasters from kinds of organisms that are being produced in hundreds or thousands of laboratories without a trace of demonstrable harm."[33] As the salvos cross each other, the lay public does well to keep in mind the advice of the U.S. Office of Technology Assessment that "no uncontestable 'scientific method' dictates which analogy is useful or acceptable."[34]

The theories of evolution are not predictive of detailed, specific pathways. Scientists are much better at ruling out what cannot happen than at predicting what will emerge from natural selection. Microbial life forms, moreover, are extremely numerous, and only a few have been subject to extensive study. These two considerations together—overly general theory and insufficient data—explain why it is easy for scientists to disagree.

Among the many unknowns is which organisms will survive under what conditions. Nor can much be said in advance about which ones will multiply where and to what degree. "Predictions of the likelihood of spread of species that have not yet been studied," Alexander asserts, "are usually highly tenuous." Sometimes even after-the-fact explanations for the survival and spread of newly introduced organisms cannot be found. With scientists having only a weak notion of how many introductions have been tried and have failed, establishing probabilities is not possible.[35]

The argument nevertheless persists that while the probability of harm coming from the introduction of engineered organisms is overwhelmingly small, "the consequences of an unlikely event," as Alexander believes, "could be enormous."[36] Davis calls this kind of thinking "pseudomathematical": since there has been "no observed formation of pathogens from crosses between non-pathogens . . . we would be extrapolating from zero." While science cannot prove negatives, other than by saying they are unlikely to occur, this is no reason, Davis insists, to stop a promising industry.[37]

In an effort to resolve this dispute, the call goes out for quantitative risk assessment. It turns out, however, that the disputants do not agree on how small is small. Ecologists refer disparagingly to such locutions as "seems very small" and "is extremely unlikely," on the ground that the purpose of risk assessment is to avoid these "kinds of qualitative and subjective judgments."[38] In response, the biologists charge that even small hazards over a large population can lead to serious damage, whereas "in modern genetic engineering not a single life has been lost; not a single person has become ill."[39] Moreover, there is no reasonable way to assign a number to the risks; the field experience of plant and animal breeders, which has engendered exceedingly few hazards, is a better guide; and breeding is done without quantitative assessment.[40]

Not everyone is agreed, then, on the desirability of field tests. Because scientists know so little, the lawyer Thomas McGarrity holds that "the test itself could present unacceptable risks." Since there is no method for labeling microorganisms, there is no way of telling whether the object of the test has moved. A great deal of research would be necessary to find this out.[41] So far there is no way to hold constant in the field such variables as soil, weather, and other populations. Robert Bohrer anticipates well the dismay of entrepreneurs and scientists when he suggests "a regulatory framework in which nothing could be done because nothing can be proven safe and nothing can be proven safe because nothing can be done."[42]

Biotechnology's proponents are in a double bind. If they cannot proceed without data and are not allowed to get data, they are finished. The microbiologists are afraid that they will not be given a chance to prove themselves, so the public will never know the good it has forgone and will never learn to laugh at its foolish fears. Hence their conclusion that they are faced with "an emotional attempt to block a new technology."[43] Between the models of fragile nature and cornucopian nature, along with the rival ways of life these models support, the twain nowadays do not often meet.

This does not mean that there is no truth, only that the situation is not set up to find it. Ideally, society would permit the planned introduction of genetically engineered organisms, allow time to pass, and then see what has happened. However, the ecologists are afraid that if the microbiologists carry out their experiments, irreversible bad effects will occur, so that the nature they have to deal with will never again be the same.

Caught between those who "would halt experiments until they are proven safe" and those who "would continue experiments until it is shown they . . . cause harm," public officials find it difficult to make intelligent decisions. The catch is that while it is possible to say that Y causes X, it is impossible to demonstrate that Y can never cause Z. As the Office of Technology Assessment maintains, "It cannot be demonstrated that recombinant DNA molecules will *never* be harmful. It can only be demonstrated that harmful events are *unlikely*. Hence, society must determine what level of uncertainty it is willing to accept."[44] That said, we leave the area of science and enter the political arena in which it is not evidence but feeling and numbers that matter.

The standard term *acceptable risk* is misleading. It suggests a chart of risks with clearly demarcated higher and lower levels. This is not possible in biotechnology. The dispute is actually over which decision rule will be safer for human life: halt until proved safe, or continue until proved harmful; in short, stop or go. The stop rule is not demonstrably safer, in this broad sense; it is one of the two rules available for choice. The benefits of a go rule might outweigh the alternative, or it might even bring about a safer society with no harm to offset. Hence, a stop rule could be more harmful by preventing results beneficial to human health. Citizens cannot just choose safety, then, since the determination of which strategy will improve safety is exactly what is at issue; there needs to be some other basis on which to make a choice.

Citizens cannot ask for more and better information, because whether or not to collect data and precisely how to gather it are part of the disagreement. The release of genetically altered organisms into the environment, after all, is a method of collecting data, as much as laboratory experiments are. Citizens are choosing the type of test along with the type of danger. If the laboratory is suspect because it is too far from "real life," and if field testing is too dangerous, a decision to ban testing means a decision not to develop biotechnology.

Anticipation versus Resilience

The criteria used in the industrialized world to achieve such extraordinary levels of health also apply to the planned introduction of engineered microorganisms into the environment.[45] One of the organizing concepts is the axiom of connectedness, whereby the good and bad consequences of technologies are inextricably bound together. Safety, therefore, has to be searched for, because the good health effects are part and parcel of the damage that accompanies them. Consider in this light the havoc that has been wreaked by organisms imported from faraway places. The healthful consequences of such importation, namely most of our food, could not have been gained without risking the bad. It is doubtful whether any set of regulations could have prevented the harmful without also keeping out much or most of the good.

Biotechnology can also learn lessons from unexpected and devastating hazards, such as AIDS. Given that bad things are bound to happen, things whose very quality, let alone whose probability, no one can now suspect, society has two strategies by which to safeguard itself: anticipation and resilience. The strategy of anticipation, or preventing bad things from happening, works only when certain conditions are met—when we know the what, when, why, and how of a threat. Otherwise, society would waste its resources on harms that would never occur. Since we seldom know what harmful events will occur, nor when, nor necessarily how to counter them, anticipation is out, lest the cure be worse than the disease.

We must turn, therefore, to the other major strategy, that of resilience, in which we seek to learn from adversity how to do better. Sometimes such a strategy is called trial and error, as opposed to the anticipatory strategy of "trial without error," because it does not provide any advance assurance that no harm will come. Resilience is best obtained by the accumulation of global resources, like knowledge and organizational capacity, that can be converted into other forms and can be directed to a wide variety of purposes. Applied to the occurrence of AIDS, resilience calls for the generation of strong and vibrant biotechnological sciences in advance of knowing exactly where they will be directed. Only in this way can society react to the wholly unexpected with any hope of success. In fact, the ability both to discover the AIDS virus so quickly and to engender multiple, international efforts to devise countermeasures depended on the prior existence of resilience provided by lots of independent trials and errors.

But it might seem that nothing could possibly be wrong with a cautious conservatism, that it is better by prevention to be safe than sorry. Without doubt, the introduction of new things, the new interaction of old things—genuine surprise—brings with it adverse effects on health, safety, and the physical environment. Yet every year or two health rates and longevity go up (see Table 6.2). Something else must be happening to counteract the effect of negative episodes. That something comes from the development of positive new global resources, such as skills and organizational capacities. In short, our health and safety are largely a function of our society's accumulation of global resources.

The rationale for accepting the harm that accompanies risktaking is better health and improved safety. People become healthier and safer in the same way that they obtain other goods of value—by trial-and-error risktaking. Trying to get safety without searching for it is like trying to discover errors without undertaking trials to uncover them. Safety without risk is a delusion that a "Safety General," if there were one, would tell us is bad for our health.

Think of opportunity risks as the probability of damage from new departures. Think of opportunity benefits as the probability of discover-

Table 6.2   Death rates for selected causes (per 100,000 population)

Cause of Death	1985	1900–1904
Typhoid fever	na	26.7
Communicable diseases of childhood	na	65.2
Measles	—	10.0
Scarlet fever	na	11.8
Whooping cough	—	10.7
Diphtheria	na	32.7
Pneumonia and influenza	27.9	184.3
Influenza	0.8	22.8
Pneumonia	27.1	161.5
Tuberculosis	0.7	184.7
Cancer	191.7	67.7
Diabetes mellitus	16.2	12.2
Major cardiovascular disease	410.7	359.5
Diseases of the heart	325.0	153.0
Cerebrovascular diseases	64.0	106.3
Nephritis and nephrosis	9.4	84.3
Syphilis	0.0	12.9
Appendicitis	0.2	9.4
Accidents, all forms	18.8	na
Motor vehicle accidents	10.6	na
All causes	874.3	1,621.6

Source: The 1988 Information Please Almanac (Boston: Houghton Mifflin, 1988), p. 793.

ing new combinations that improve safety. As the economy develops, opportunity benefits grow faster than opportunity risks. The danger of low-probability but high-damage risks is overcome by two factors. First, low probability often means just that: the feared catastrophe will fail to occur. Second, great advances do occur, as with public health measures or antibiotics, and opportunity benefits slowly but steadily accumulate.

Placing the burden of proof on new things ignores the likelihood that most of the dangers guarded against will not in fact materialize. Hence, attempting to prevent all dangers imposes unnecessary health and safety costs on society. Overusing preventive measures uses up huge amounts of resources in a vain effort to circumvent the principle of uncertainty, which warns us that we will not guess right most of the time. Placing the burden of proof on regulation, or allowing new things to take place unless there is reason to believe regulation will do more good than harm, preserves resources that will enhance our capacity to deal with hazards as they become known. Thus, reducing society's wealth decreases the health and safety of its people.

The anticipatory strategy ignores the fact that preventing progress can hurt us. Society cannot continue to achieve the past century's advances in health and safety without continuing to encourage innovation. The more society limits trial and error, the less innovation there will be, the fewer general resources will be generated, and the sicker people will be. Old hazards will not be reduced as much as they would have been. New benefits, such as those from biotechnology, will be much delayed or may never materialize. Society's capacity to synthesize responses to emergent dangers no one ever thought of before will be impeded. We cannot have the one, namely steady improvement in health and safety, without the other, namely an innovative technology based on rapid, continuous, noncentralized, trial-and-error decision making.

The proper criterion of choice is net benefit. If we hinder progress on the ground that it brings some "bads," we will deny ourselves even greater "goods." If it seems too cruel to contemplate any harm at all, the even greater cruelty is to abandon the net benefit, for giving it up guarantees that more people will have worse health.

In cultural terms, the resilient position is largely individualist, placing fewer impediments in the way of trial-and-error learning, but it is also partly hierarchical, placing greater weight on knowledge validated by the microbiologists who know most about this subject. Neither side has to be right. The difficulty for the nonscientists who make the laws is that each side appears plausible.

The problem presented by the efforts to commercialize engineered organisms is where to place one's bets: whether to achieve superior safety by regulatory delay or by rapid progress, maintaining only whatever regulation or surveillance is compatible with that progress.[46] The benefits of genetic engineering are acknowledged on all sides to be immense. They run from vast amelioration or avoidance of existing illnesses to enormous improvement in the ability to produce inexpensive food. Nothing bad has happened yet. But it might. The question is how best to protect ourselves, and against what.

The question of allowing planned introductions of engineered organisms involves hazards prevented but also possible health gains lost. If we go ahead, we will have expanded our knowledge and be better prepared for whatever comes. The loss of safety due to inaction is important. The more likely dangers are those, such as new diseases, that will have nothing to do with planned introductions. Should global warming —or for that matter freezing—materialize, for instance, declines in food production could well be made up by genetic engineering.

Moreover, if there is to be stringent regulation, it is more appropriate for less developed countries than for the United States. Trials, and thus possibly errors, ought to take place in countries with the best science and scientists.

Politics

Public opinion about genetic engineering is unsettled. Though only one-fifth of Americans surveyed have heard of the dangers, half think we are in for trouble. The vast bulk of the population believes biotechnology will do more good than harm, but is still leery of it. One-third think it would be just as well if we did not know how to manipulate genes. There is a positive attitude toward tests of genetically altered organisms but a negative attitude toward large-scale applications. The more the risks remain unknown, the less the public is inclined to accept them. A strong majority favors proceeding with technological development but under strict regulation. With opinion divided between increased control and the status quo, reducing regulation is unlikely.[47]

Because the procedures involved in biotechnology are fairly simple and require relatively small investments, because the tax laws are favorable, and because Americans may be more entrepreneurial than other nationalities, almost all of the small start-up companies involved in the commercial applications of biotechnology, with the exception of a few in Western Europe, are in the United States.[48] These ventures often involve a collaboration with university scientists. Whereas in some countries closer relations are sought between universities and industry, in the United States alarm has arisen about the possible corruption of the campuses by commerce.[49]

Because the system bias favors big business as distinguished from small, strict regulation is bound to favor bigness, for only large firms can wait out years of delay or pay specialists for help in getting through the system. Yet ecologists maintain that industry, especially the small entrepreneur, "not infrequently puts profit before protecting the public or even considering the public good."[50] When enterprises first have to prove that they operate according to the state's idea of the public interest, the system is called *socialism*. Profit is not in good odor. Edward Yoxen sees in biotechnology "the power to exploit the genetic resources of plants in order to gain control over future markets." Another term for this exploitation is *capitalism*. Here the exploitation of nature is directly comparable to the exploitation of people. Individualists are for and egalitarians are against planned introductions because the former see the economic system as beneficial, while the latter believe it brings harm. Yoxen fears "a new structure of dependence on agribusiness," the "savage irony" of which is that "many of the genes used to create the new plants will have come from those very countries where food is short. The genetic resources of such countries, which form a large part of the world's ever-decreasing number of species, will be used abroad and patented, or offered back to them in new plants, at prices they can ill afford."[51] Although the economic value of those genes is zero—or close to it—unless and until they are put to use, this kind of rhetoric, which stresses the unfair subordination of the weak by the strong, pervades the discussion of commercialization of genetic engineering.

If one believes that the social relations sponsored by industry are bad for public health, one would not want engineered organisms to become a prop to an economy based on toxic chemicals by increasing the resistance of crops to herbicides. Thus, Rebecca Goldburg argues that "scarce research dollars should be invested to develop alternatives to pollutants—such as biological control—not ways to make pollutants economically less damaging."[52] Accordingly, environmental groups supported the application of a biotechnology company to try out a gene for a bacterial protein that could be introduced into corn so as to make it lethal to the corn borer, at the same time that Jeremy Rifkin wanted tests to show that the new product would not also kill pests that kill weeds.[53] Whereas concern over pesticides has led to the development of insect-resistant crops, these plants pose a problem in that they may protect themselves by greatly increasing, in the food, the natural poisons they use to ward off predators. In the case of a new variety of celery, the substantial increase in its "natural" pesticides led to its rejection.

A good test of whether opposition to genetic engineering is based on fear of its effects or on opposition to capitalism is the development of bovine somatotrophin (BST), also called bovine growth hormone, which promises to increase the production of milk per cow by up to 40 percent with no increase in cost. One might think that such a development of cheaper, nutritious food would be greeted with acclaim. Not so. Critics complain that BST would flood an already saturated milk market, drive dairy farmers out of business, and cost a fortune in price supports.[54] The days when it was thought that inexpensive food is a boon, especially to poor people, seem to be gone.

Regulation abroad of the introduction of engineered organisms varies from the five-year prohibition secured by the Green party in West Germany, to fairly strict controls in Britain, to minimum regulation in Italy.[55] The European Community is considering a moratorium on planned introductions. No one can yet say that competitiveness in the United States will suffer because of existing regulations. But these regulations have slowed it down.

I have reached three conclusions about the politics of genetic engineering. First, academics have been split off from industrial biologists. When it looked as though research would be banned, the scientific community banded together to preserve its prerogatives. Once research left the academic laboratory, however, entrepreneurial biologists were left to fend for themselves as if contaminated by capitalism.

Second, while efforts at regulatory control are likely to be tried, they are even more likely to fail. For one thing, the speed of discovery is so great that regulation cannot keep up with it. There are too many holes in too many dikes for a regulator to keep a finger in all or most of them.

Third, direct action may be able to stop the introduction of engineered organisms. If consumers can be made sufficiently fearful, they will refuse to buy products produced in this way. Consequently, research, without a practical outlet, will atrophy.

The real choice may be between setting a brisk pace in the United States, where communication and control are apt to be at their best, or doing things more haphazardly under worse conditions in poor countries, where significant errors are more likely to be hidden. The divisions among American scientists, based on conflicts among rival ways of life, can create sufficient doubt for the momentum of development to move elsewhere. But the gene (if not the genie) is out of the laboratory for good.

Everyone agrees on minimal regulation so as to prevent known pathogens from being used. Beyond that minimum, regulatory efforts can be expected to intensify. The administrative machinery is geared up, and there are many aspirants. More important, the causes of the differences over policy lie deep in a distrust of institutions, with those people who believe in greater equality of condition more distrustful of established authority, even in science, than are the more entrepreneurial and hence more science-loving individualists. As political divisions over equality grow, so will disputes over regulation.

Although optimistic microbiologists believe that they will not be hampered further by regulation, I think otherwise. The Ecological Society of America (ESA) has tried in a report to defuse the controversy over engineered organisms by moderating its demands for regulation.[56] But it has not dropped a single one of the old criticisms and complaints. Its demands for regulation are no less stringent than before. These requirements cannot be met. Discussion of the ESA position is included here in the section on politics because I interpret it as a bid by ecologists to gain legitimacy for intervention.

Appreciating the benefits that engineered organisms may bring, the ESA defines its guiding principle as that of encouraging "innovation without compromising sound environmental management." The terms the ESA uses, however—*careful design, proper planning, regulatory oversight, minimal ecological risk*—suggest otherwise. Few projects can meet their criteria. The ESA denies that "precise genetic alterations ensure that all ecologically important aspects of the phenotype can be predicted for the environments into which an organism will be released." But no one could achieve the miracle of predicting "all important aspects" of anything. The ESA denies "that barriers crossed by modern biotechnology are comparable to those constantly crossed in nature." It therefore advocates caution in the form of small field tests "under appropriate regulatory oversight"—*oversight* being the same term used to speak indirectly of congressional control over the executive branch.[57]

To define what measure of regulation is appropriate, the ESA asserts that "the absence of an immediate negative effect does not ensure that no effect will ever occur." Thus, regulation is forever. The ESA further argues that "an overall record of little or no hazard stemming from the release of the products of traditional agricultural breeding does not legitimately warrant exemption from oversight for future introductions of transgenic organisms that these traditional techniques could not have produced." Thus, evidence of no harm gains no exemption from regulation. Although in principle the ESA might contemplate removing classes of experiments from regulation, it "cannot now recommend the complete exemption of specific organisms or traits from regulatory oversight." Instead of relief from regulation, the ESA offers "to assist in the development of methods for scaling the level of oversight appropriate to individual cases according to objective scientific criteria."[58] Thus, the consent of ecologists must be given before even individual research projects will have their regulatory burden reduced. This is not peace but war under another name.[59]

To begin with what is agreed about genetic engineering, the likely benefits of its products are incalculably large, while the potential hazards, though the form they might take is uncertain, are extremely unlikely to materialize. The catch is that while the dangers are indeed improbable, they just might be exceedingly large. What to do?

If the regulatory standard of "trial without error," or no trials without prior guarantees against harm, is followed, no dispersal of genetically engineered microorganisms into the environment is permissible. Because this stringent criterion of choice—no harm, however improbable—violates the criterion of net benefit, the result is likely to be more people who are sicker and hungrier and otherwise afflicted with worse health than they had to be. Milder forms of regulation have the advantage of being more acceptable to the regulated, but they have the disadvantage of not really satisfying those who do not trust institutions to do the job.

Stringent regulation, displacing planned introductions of genetically altered microorganisms from better places to worse, will encourage escape from its restrictions. Regulation therefore will not be effective. If regulation will not work, and if free development is unacceptable to politicians and the public, something else has to be done.

The problem has been misconceived as one of limitation of the extent of regulation, when it ought to be thought of as one of development. Since introductions into the environment will take place, especially as research proceeds by leaps and bounds, it is futile to attempt to stop or slow down what every interested party around the world has the resources and the know-how to do. When it is possible quickly to make millions of copies of a gene, when gene transfer to plants is "highly developed, easy to use, and applicable to a large number of . . . species," and when any graduate student can do it, regulation can be only partially effective.[60] Worse still, regulation can be counterproductive if it leads to the erroneous conclusion that people know what is and is not happening. Since the problem in keeping the prince of health-through-biotechnology from becoming a monstrous frog is residual uncertainty, the better course is to force that uncertainty into the open by pushing the phase of development in such a way as to encourage maximum disclosure.

Disclosure will have obstacles to overcome. One is the disclosure of proprietary information. Here the competition fostered by openness can be expected to make up for any losses due to greater disclosure. The more difficult problem is the damage done by reports of harmful results. A backlash would doom the product. To look at the matter another way, if sanctions are placed on the appearance of bad news, incentives will be created to hide the evidence. This is the last thing we want. By pursuing open research and by reducing incentives to keep research secret, we increase the likelihood of learning what, if anything, has gone wrong.

In order to bring harmful effects out into the open, incentives could also be given to those who find harmful genetically engineered microorganisms in the field. Prizes might be awarded to those who can create monsters in the laboratory. The very suggestion calls attention to the difficulty of the task.

The best policy is "full speed ahead" with the introduction of genetically engineered organisms into the environment. The net benefit to public health promises to be huge, for discoveries benefiting health and safety are almost certain to be made a lot faster than hazards materialize. And to minimize those hazards, measures should be undertaken to drive harmful effects, if any, to the surface, so that they can be recognized, appraised, and dealt with.

Conclusion

The basis of the disagreement between those who wish to say yes to the introduction of genetically engineered organisms into the environment and those who wish to say no comes down largely to their preference for different ways of life, or cultures. Were this factor not operating, all the other differences would count for much less. As a result of this factor, two predictions are possible. First, the disputants will not be able to agree on the facts about planned introduction, the risks, and therefore the proper extent of governmental regulation. Cultural conflict will reinforce self-interest. The microbiologists who do the research and the businesses that attempt to market the results will oppose restrictions on their activities, while the environmentalists, bolstered by the ecologists, will seek increasingly restrictive regulation, up to and including lengthy moratoriums. Entrepreneurs and researchers who cannot afford the lengthy delays will have to leave the field or go to third-world countries. Efforts to reason together will not succeed.

For the two sides to come together, the more extreme critics, such as the Greens, would have to be convinced that use of genetically altered organisms would lead to greater equality of condition in society, including diminishing the differences between scientists and nonscientists. Until they are so persuaded, they will continue to oppose a technology that they believe to be evil to the nth power. They believe it is allied with those who profit at the expense of other people's health. They believe that the hubris of so-called experts gives these scientists the unholy power to experiment secretly with the lives of ordinary people. The profits, in their estimation, make the rich richer, while the family farmer and the third-world peasant are made poorer. Worst of all, unnatural, artificial life forms are foisted on an unwilling people, replacing the wholesome variety that nature has passed down to us.

The second prediction is that now that the critics have found their "good" scientists, the ecologists, these scientists will be increasingly required as members of any "representative" panel of scientists considering regulation, until their consent becomes essential for any introduction to take place. Instead of entire classes of experiments being exempted from regulation, as the microbiologists had hoped, case-by-case approval by expanded committees will be required. This trend will reduce the large disparity of status and grants between microbiologists and ecologists and will increase their equality of condition.

The drive to maintain economic competitiveness, which supporters of biotechnology hope will come to their rescue, is by no means automatic. Critics can leap over this obstacle through international regulations that halt or confine introductions of genetically engineered organisms, or they can create such anxiety that few will be willing to buy such products or even to tolerate such research. Thus, the Common Market might refuse products of biotechnology as inherently polluted. Or American supermarkets might continue to refuse to sell certain products, such as milk coming from cows using genetically engineered organisms—not because anything is wrong with the product but because the slightest suggestion of taint reduces sales.[61]

In view of these predictions, the usual optimism of microbiologists may be misplaced. Just because, in this instance, doing well also means doing good and the science seems to be on their side, it does not follow that the microbiologists will prevail. Cultural conflict is hardball. Unless they work a lot harder at persuading the public and the politicians, they will likely lose. And with them, for a time, will go the good they have done and the much greater good their branch of science might have done.

The difficulty of prediction lies in the existence of two trends leading in opposite directions. The preventive or anticipatory trend might impose regulations so stringent that introductions of genetically engineered organisms are forbidden, or so severely limited that virtually no experience is gained. Also, public reaction might be so negative that no products can be successfully marketed. Consequently, neither the promise nor the peril of genetically engineered organisms would be better understood. Conversely, the resilient trend might manifest itself if such significant advances occur that public opinion and commercial considerations coincide to push ahead the frontiers. The promise of thousands of applications might then lead to general rules for expediting them. But neither prospect is likely in the United States; consequently, trials might simply be shifted to newly industrializing countries.

Conclusive experiments, acceptable to all sides, should be tried so that introductions can proceed under greater or lesser degrees of regulation. The results, however, would probably not meet universal acclaim, as there are always site-specific conditions, such as the process or organism involved and the state of the environment, that would invalidate the results in someone's view. The resilient path will prevail, but not by persuading the proponents of prevention; it will prevail, I expect, because prevention either proves infeasible or is overtaken by new circumstances requiring solutions that only biotechnology can provide.

A race is being run whose outcome is uncertain. Whether, as Harvey Brooks puts it, science proves to be a delicate flower, easily uprooted, or a sturdy weed, with roots so tenacious they can be torn but not destroyed, is now being decided. I think the flower will bloom eventually. But eventually is a long time, and much of value to human life, not only health and safety but also the means of future improvement, may be long delayed.

Notes

1. See R. Heiner, "The Origin of Predictable Behavior," *Am. Econ. Rev.* 13, no. 4 (1983): 560–95.

2. M. Schwarz and M. Thompson, *Divided We Stand* (Hertfordshire: Harvester-Wheatsheaf, 1990).

3. "MIT Closes 'Biotech' Department," *Bio/Technology* 6 (February 1988): 108.

4. R. W. F. Hardy and D. J. Glass, "Our Investment: What Is at Stake," *Iss. Sci. Technol.* 1, no. 3 (1985): 69–82.

5. In humans, genetic engineering in germ line cells is distinguished from gene therapy in somatic cells. Successful or not, gene therapy does not alter the ability of the host to pass on traits to children; only the host's body is affected. Though this interference with the body worries many people, it is sufficiently limited in scope not to affect the larger problems of regulation.

6. A government report observes, "Altering the hereditary characteristics of an organism by using rDNA is just one of the several methods of genetic engineering. The definition of rDNA refers specifically to the combination of the DNA from two organisms *outside* the cell. If the DNA is combined *within* living cells, the Guidelines do not pertain. [There are] several methods that achieve the same goal—transferring genetic material from one cell to another, bypassing the normal sexual mechanisms of mating. It is particularly significant that DNA *from different species* can be combined by all these mechanisms, only one of which is rDNA." Office of Technology Assessment, *Impacts of Applied Genetics: Micro-organisms, Plants, and Animals* (Washington, D.C.: U.S. Govern-

ment Printing Office, 1981). See also R. A. Zilinskas and B. K. Zimmerman, *The Gene-splicing Wars* (New York: Macmillan, 1986).

7. M. Douglas and A. Wildavsky, *Risk and Culture* (Berkeley: University of California Press, 1982); M. Thompson and A. Wildavsky, "A Cultural Theory of Information Bias in Organizations," *Journal of Management Studies* 23, no. 3 (1986): 273–86; M. Thompson, R. Ellis, and A. Wildavsky, *Cultural Theory* (Boulder: Westview Press, 1990); K. Dake and A. Wildavsky, "Theories of Risk Perception: Who Fears What and Why?" *Daedalus* 119, no. 4 (1990): 41–60.

8. D. Bloor, "Polyhedra and the Abominations of Leviticus: Cognitive Styles in Mathematics," in *Essays in the Sociology of Perception*, ed. M. Douglas (London: Routledge and Kegan Paul, 1982), p. 198.

9. Scientists may well object that the better way, and for them the only valid way, is to test hypotheses, replacing a worse theory with a better one. Yet the matter in dispute is not the science, but the legitimacy of institutions. See Douglas and Wildavsky, *Risk and Culture.*

10. See M. Thompson, "Socially Viable Ideas of Nature: A Cultural Hypothesis," in *Nature, Culture, Technology: Towards a New Conceptual Framework*, ed. E. Baark and U. Svedin (London: Macmillan, 1988); Thompson, Ellis, and Wildavsky, *Cultural Theory.*

11. B. D. Davis, "Microbiology, Ecology, and Engineered Bacteria," *The Collecting Net* (Marine Biological Laboratory), August 1987, p. 13.

12. G. B. Campbell, "Biotechnology: An Introduction" (American Council on Science and Health, January 1988), pp. 5, 24, 26.

13. W. J. Brill, Letter to the Editor, *Science* 229, no. 4709 (1985): 113.

14. National Academy of Sciences, *Introduction of Recombinant DNA–Engineered Organisms into the Environment: Key Issues* (Washington, D.C.: National Academy Press, 1987), pp. 13–14.

15. B. D. Davis, "Bacterial Domestication: Underlying Assumptions," *Science* 235 (13 March 1987): 1334.

16. B. D. Davis, "Domesticated Bacteria or Andromeda Strains," *BioEssays* 7, no. 2 (August 1987): 87.

17. Davis, "Microbiology, Ecology, and Engineered Bacteria," p. 13.

18. Brill, Letter to the Editor, p. 115.

19. NAS, *Introduction of Recombinant DNA–Engineered Organisms*, pp. 14–15.

20. W. Szybalski, Letter to the Editor, *Science* 229, no. 4709 (1985): 115.

21. Davis, "Domesticated Bacteria," p. 87.

22. Szybalski, Letter to the Editor, p. 115.

23. C. Grobstein, "Future Concerns for the Scientist," in *From Research to Revolution*, ed. R. A. Bahrer (Littleton, Colo: Fred. B. Rothman, 1987), pp. 85–91.

24. Campbell, "Biotechnology," pp. 26–27.

25. Davis, "Domesticated Bacteria," p. 87.

26. Davis, "Bacterial Domestication," p. 1333.

27. M. Alexander, "Ecological Consequences: Reducing the Uncertainties," *Issues in Science and Technology* 1, no. 3 (1985): 63ff. See also in same issue Hardy and Glass, "Our Investment," pp. 69–82; T. McGarrity, "Regulat-

ing Biotechnology," pp. 40–56; "Prologue: Genetic Engineering: Balancing Risk and Reward," p. 39; S. Olson, *Biotechnology: An Industry Comes of Age* (Washington, D.C.: National Academy Press, 1986), p. 61.

28.   Alexander, "Ecological Consequences," p. 60.

29.   Olson, *Biotechnology,* pp. 60–61.

30.   F. E. Sharples, "Regulation of Products from Biotechnology," *Science* 235 (13 March 1987): 1329–31.

31.   R. Colwell, E. Norse, D. Pimental, F. Sharples, and D. Simberloff, Letter to the Editor, *Science* 229 (12 July 1985): 111.

32.   Alexander, "Ecological Consequences," p. 62.

33.   Davis, "Bacterial Domestication," p. 1335.

34.   OTA, *Impacts of Applied Genetics,* p. 200.

35.   Alexander, "Ecological Consequences," pp. 60–61; Olson, *Biotechnology,* pp. 58–59.

36.   Alexander, "Ecological Consequences," p. 64.

37.   Davis, "Microbiology, Ecology, and Engineered Bacteria," p. 14

38.   Colwell et al., Letter to the Editor, p. 111. Rough quantities, like "more or less," are better than none. See also E. Nichols and A. Wildavsky, "Regulating by the Numbers: Probabilistic Risk Assessment and Nuclear Power," *Eval. Rev.* 12, no. 5 (1988): 528–46.

39.   Szybalski, Letter to the Editor, p. 112.

40.   Brill, Letter to the Editor, p. 116; W. J. Brill, "Safety Concerns and Genetic Engineering in Agriculture," *Science* 25 (January 1985): 381–84.

41.   McGarrity, "Regulating Biotechnology," pp. 45–46.

42.   R. A. Bohrer, "The Future Regulation of Biotechnology," in *From Research to Revolution,* p. 112. (In fact, genetic markers are used to label the progeny of a bacterial strain.)

43.   Campbell, "Biotechnology," p. 27.

44.   OTA, *Impacts of Applied Genetics,* pp. 203–4.

45.   A. Wildavsky, *Searching for Safety* (New Brunswick, N.J.: Transaction Books, 1988).

46.   See E. Bardach and R. A. Kagan, eds., *Social Regulations: Strategies for Reform* (San Francisco: Institute for Contemporary Studies, 1982).

47.   Office of Technology Assessment, *New Developments in Biotechnology—Background Paper: Public Perceptions of Biotechnology,* OTA-BP-BA-45 (Washington, D.C.: U.S. Government Printing Office, 1987), pp. 32ff.

48.   Olson, *Biotechnology,* pp. 85–86.

49.   J. Johnson, "Biotechnology: Commercialization of Academic Research," Congressional Research Service, Issue Brief (7 February 1984), pp. 5ff.

50.   Alexander, "Ecological Consequences."

51.   E. Yoxen, *The Gene Business: Who Should Control Biotechnology* (New York: Harper and Row, 1983), pp. 121–22.

52.   R. Goldburg, "An Environmentalist's View," *Bio/Technology* 6 (March 1988): 336.

53.   J. Kwitny, "Crop Genetics' Bid to Test Corn Strain Creates Rift in Environmental Circles," *Wall Street Journal,* 25 March 1988, p. 6.

54.   G. S. Becker and S. Taylor, "Bovine Growth Hormone (Somatotropin):

Agricultural and Regulatory Issues," Congressional Research Service, 86-1020ENR/SPR (20 November 1986), pp. 1–3, 24–26.

55.   P. Newmark, "Discord and Harmony in Europe," *Bio/Technology* 5 (December 1987): 1281–83.

56.   J. M. Tiedje, R. K. Colwell, Y. L. Grossman, R. E. Hodson, R. E. Lenski, R. N. Mack, and P. J. Regal, "The Planned Introduction of Genetically Engineered Organisms: Ecological Considerations and Recommendations," *Ecology* 70, no. 2 (1989): 298–315.

57.   Ibid., pp. 299, 301, 302, 304; A. MacMahon, "Congressional Oversight of Administration: The Power of the Purse," *Poli. Sci. Q.* 58 (June 1943): 161–90; (September 1943): 380–414.

58.   Tiedje et al., "Planned Introduction," pp. 305, 306, 307.

59.   The closest the ESA comes to specifying criteria for reduced risk experiments is as follows: "By engineering a stable construct, verifying its stability in the presence of known vectors, controlling the densities of introduced organisms, and characterizing potential environments that will receive the transgenic organisms, field experiments with low risk for gene exchange are possible" (ibid., p. 304). Who will be the first to characterize (all?) potential environments or verify stability (under all possible conditions)?

60.   R. B. Goldberg, "Plants: Novel Developmental Processes," *Science* 240 (10 June 1988): 1465.

61.   See B. Richards, "Sour Reception Greets Milk Hormone Mixing: Food Biotechnology Faces Resistance," *Wall Street Journal*, 15 September 1989, p. B1. Four of our country's largest chains will not carry milk made with the use of bovine growth hormone. Safeway said it did not want to be a guinea pig. Kraft said its products were used by small children and "just mentioning hormone additives in milk repels many consumers." Consumers view drug and chemical companies as "the archvillains of the food world." Whatever these companies do, a marketing strategist advises them never to refer to "genetic engineering, as the phrase engenders negative reactions." Time will tell whether the companies' reliance on the Food and Drug Administration and scientists to reassure the public will work.

7

Plant Agriculture

Charles J. Arntzen

The use of biotechnology in crop improvement is as old as agrarian civilization. Primitive societies selected seeds for later planting and experimented with new crops. In the twentieth century successive technological advances have dramatically increased crop yields. This chapter will describe first the role that classical genetics has played in these advances, then the development of a system of intensive agriculture, and finally the new technologies provided by plant cell culture and molecular genetics. Problems faced by contemporary agriculture will also be discussed, including the conflict between the demands of environmental protection and the pressures of population growth and increased affluence.

Crop Genetics

The greatest success of classical genetics in the directed manipulation of genetic systems was the production of hybrid seed. This process involves selection of inbred male and female parents that are themselves not superior, but whose mating gives uniform and excellent progeny.

The vast majority of all field corn produced around the world is now grown from hybrid seed, developed commercially and sold to the farmer for each year's crop. Its production is conceptually simple. Inbred parental lines are grown together in the same field. The male (tassel) and female (silk) flowers are on different parts of the same plant. The tassel is mechanically removed early from what is to be the female parent, which later becomes cross-pollinated from a plant growing nearby. Hybrid seed is then easily collected from female (emasculated) plants. The plants from these seeds express hybrid vigor—that is, they grow faster and yield more than either of the parental lines.

Similar manipulation has been implemented for other crops, such as rice, on the basis of increased understanding of their genetics. Because the flowers of rice, containing both male and female parts, are very small, the manual emasculation used in corn is impractical on an agricultural scale. Fortunately, in 1970 a Chinese geneticist discovered a wild rice plant that was male-sterile; that is, it lacked the ability to produce pollen. By careful hand pollination, using pollen from a cultivated rice plant, a few precious seeds were obtained. Over the last two decades their progeny have been used extensively to introduce the male-sterility trait in various breeding lines of rice, which have now been adopted for commercial production. China now leads all nations in hybrid rice research and production, and the technology is being used in research institutes around the world. This is only one example of the introduction of "exotic" genetic material into important crops via classical biotechnology.

Intensive Agriculture

The spectacular increase in global food production over the past fifty years is due to advances not only in crop genetics but also in several other technologies, the ensemble constituting intensive agriculture. Among the additional technologies, agricultural mechanization greatly reduces the need for human labor and makes the conditions of crop production more uniform; improvements in irrigation increase the productivity of crops in dry-land areas of the world; chemical fertilizers supplement the naturally available minerals in the soil; and crop protection chemicals control competition by weeds and attack by insects and by microorganisms. Intensive agriculture is increasingly being recognized as an integrated science; that is, its components are highly interactive.

Intensive agriculture, especially for rice, has probably been used longer in China than in any other country. Techniques to improve crop productivity have been studied and recorded in China for nearly five hundred years. For example, in the seventeenth century *Shen's Agricultural Book* described top-dressing rice plants with a manure in accordance with the growth of the plant. The same principle was utilized by American Indians when they added a dead fish to the soil when they planted seeds of corn. These early agriculturalists recognized what was later understood as the need for additional nitrogen supply.

In the twentieth century, synthetic nitrogen fertilizers became available with steadily decreasing cost per unit. But their adoption has not always been simple, and it illustrates the need for interaction among the components of intensive agriculture. In most traditional cereal crops, which show the greatest response of yield, nitrogen fertilizers also cause the stems to elongate and hence to topple or break in heavy winds. The grain is then difficult or impossible to harvest, and it is more susceptible to disease.

During the 1950s and 1960s research focused on genetic selection for shorter, stiff-stalk varieties of wheat and rice, and the resulting dwarf types have now been accepted around the world. Varieties such as these, used in conjunction with fertilizers and other inputs in an integrated technology, have resulted in the dramatic yield increases now referred to as the green revolution.

Mexican wheat yields are an example of the success of integrated technology. In 1945 these yields were low and stagnant, at 750 kilograms per hectare, and Mexico produced only 40 percent of its total wheat consumed. By 1984 the average yield had reached 4,300 kilograms per hectare. Similar dramatic increases in rice production have been observed in China, India, Pakistan, and other nations that adopted the integrated technologies of the green revolution.

Intensive agriculture is now practiced in all developed countries. The practice varies among crops and among regions of the world, but it is always focused on one primary objective: maximum yield of food and fiber. One reason is the need to enhance profitability for the farmer. Another reason is the need for food production to match an expanding world population. The world population exceeds 5 billion today, and since it is expanding at the rate of about 84 million people annually, it is expected to reach 6 billion by the turn of the century. Moreover, between 600 million and 800 million people already live in abject poverty. From these statistics it is clear that by necessity agriculture will become even more intensive as we approach the next century.

The future world demand for food supplies will be determined by the growth of both population and per capita food consumption. In fully developed countries, such as the United States, Canada, and those of Western Europe, both population and consumption growth have slowed and can be expected to remain steady. But while population is also growing slowly in Eastern Europe, the Soviet Union, and the Pacific Rim, per capita food consumption, especially of meat products, is rising fast in those parts of the world. This nutritional change will require substantially higher supplies of grain for livestock feed.

In many developing countries, however, the population is continuing to grow rapidly and will require even greater increases in food supply. Moreover, malnutrition is widespread in many poor nations. By the turn of the century it is anticipated that total annual food supplies in the developing world will have to increase by nearly 1 billion tons. In addition, economic development in the poor countries will increase their demand for poultry and livestock products, which they now consume in very small quantities. Much larger quantities of feedgrain will then be needed.

Societal Perceptions

The integrated use of the various new technologies has resulted in a stable supply of low-cost, abundant, and healthy food in the developed world. As might be expected in a period of well-being, the technologies have been scrutinized for negative as well as positive effects. Perhaps the single largest issue in the United States and Europe is public concern about the presumably unavoidable adverse effects of agricultural chemicals on the environment, especially effects involving public health and the survival of wildlife.

Coming to grips with societal perceptions of the new agricultural technologies will be a complex problem. Clearly, we cannot ignore the close tie between technologies and levels of food production. Agricultural technologies of the 1930s, which were adequate to feed a world of 2 billion people, would have been unable to produce the food needed for the current population of 5 billion. And existing technologies will not be adequate to feed the 6 billion people expected at the turn of the century, let alone the more than 8 billion who will inhabit the planet in the year 2030. A risk-benefit analysis of future developments in agricultural technology must take these principles into account.

The task facing agriculturists over the coming years, as we enter a period of the most rapid population expansion this globe has ever witnessed, will be monumental. In addition to the pressures to increase the economic return to the family production units in all countries, there are needs to find ways to spare the environment from potential negative impacts of agricultural technologies, and there is a rising scientific and public expectation that the solutions will come largely from recent breakthroughs in biotechnology.

"Modern" crop biotechnology is derived from traditional crop breeding, but with a much greater emphasis on cell and molecular biology. Contemporary plant biotechnology has gone through two distinct but overlapping phases. In the first, lasting several decades, methods were created to isolate plant cells, to culture them much like bacteria, and to regenerate complete plants. In the second phase, which began in the 1980s, techniques were developed for the addition of DNA (genes) directly to plant cells, giving rise to transgenic plants.

Cell and Tissue Culture

After it was shown that plant cells can be cultured and can regenerate an intact plant, several practical applications of the technology emerged, which are now standard horticultural and agronomic practice. One major use is the clonal (nonsexual) generation of a large number of identical individuals from a single tissue sample selected for a superior trait. For example, many virus-free horticultural and vegetable strains are now routinely produced by isolation and clonal regeneration of disease-free cells from plant and shoot tips.

In addition, plant cells and tissues in culture frequently undergo a process of mutational change and chromosomal rearrangement that gives rise to new genotypes. This process is now widely utilized to generate variants that are propagated for practical use. A downside of this approach is that the outcomes are almost entirely empirical; one cannot predict in advance the character of the new phenotype. Nevertheless, the approach has generated valuable genetic variants in horticultural varieties.

Gene Transfer

Recombinant DNA technology has increased the speed and efficiency of crop improvement. It allows the genetic transformation of plants by the introduction of specific foreign DNA with predictable effects. The procedure is rapidly becoming a useful addition to crop breeding.

The best-developed method for introducing DNA into plants uses the soil bacterium *Agrobacterium tumefaciens*. In nature this bacterium causes a disease of certain plants, called crown gall, in which a localized site of uncontrolled cell divisions yields tumors that are called galls. The mechanism is the transfer of a particular region of DNA, called T-DNA, from the bacterium into the plant cells, where it is incorporated into the chromosome; even if the bacterial infection then ceases, symptoms still develop. A small, specific sequence of the T-DNA causes the infected plant to synthesize plant hormones, and the resulting imbalance in the developmental program of the plant tissue gives rise to the characteristic tumorlike gall.

The tumor-inducing mechanism of the bacterium has proved useful for genetic engineering. Studies in the early 1980s showed that T-DNA can have almost all of its middle portion removed, and genes from other organisms can be inserted, without affecting the transfer of this DNA from Agrobacterium to plant cells. (These modified portions of T-DNA are termed vectors, since they provide a mechanism to shuttle new genes into plants.) Improved T-DNA vectors were constructed containing convenient restriction endonuclease–sensitive sites for inserting or excising a gene or DNA fragment, and also containing a gene whose product easily identified plants or cells that have received the T-DNA. In addition, the vectors were "disarmed" by removing the tumor-causing genes, so they can mediate gene transfer to plant cells without causing disease.

Agrobacterium-mediated gene delivery was originally limited to relatively few plant species, such as the tobacco plant, that are susceptible to infection by this microbe. Discovery of ways to modulate the host range of the bacterium resulted in the extension of this technique of gene transfer to a wide range of crop plants. Unfortunately, the system has not yet proved to be always reliable and reproducible, especially with monocotyledon species, including corn and most cereals. As a result, alternative strategies are being developed for DNA delivery to some crop plants.

Recently, physical methods of plant cell transformation have been developed. They have the advantage that genes of interest can be inserted into plant cells without any companion DNA derived from bacteria. One technique, called particle bombardment, uses a modified rifle shell casing packed with explosive powder and DNA-coated particles of metallic material a few microns in size. Firing of the cartridge propels the particles into the plant tissue. The release of the DNA then causes the genetic transformation of some cells. In another approach DNA is directly injected into plant cells, using a very thin glass tube.

With the advent of new genetic vectors, other DNA delivery methods, and methods to identify transgenic plants, nearly three dozen species of crop plants have been successfully transformed. These include rice, corn, soybeans, cotton, and potatoes. Within a few years techniques will have been developed for the genetic transformation of all major crop species.

Gene Regulation

Transformation with Agrobacterium vectors in plant cells was first detected by the expression of a bacterial gene encoding resistance to the antibiotic kanamycin. Subsequently, other plant and animal genes that endow the resultant plant with more interesting traits were introduced.

Moreover, each cell, tissue, organ, and developmental stage (such as the formation of leaves as opposed to flowers) has characteristic patterns of gene expression, and vectors have been created that can dictate the site and timing for expression of the genes that they introduce.

Eventually it will be possible to produce transgenic crops in which the novel genes are expressed only in cells or tissues where there is some benefit. For example, it may be possible to control the boll weevil by producing biological insecticidal agents only in the fruiting body of a cotton plant, but not in other plant parts where there could be negative effects on beneficial insects. Similarly, gene products that enhance nutritional value may be targeted to the fruit, seed, or other harvested part of food crops. An example would be corn kernels enriched in certain amino acids that are essential in human and animal nutrition.

Herbicide Resistance

The principle outcome of plant genetic engineering has been a rapid expansion in the understanding of plant cell biology, genetics, and molecular controls of form and function. This understanding is now helping to solve practical problems on behalf of the farmer or food consumer. One of the first projects was to understand the mechanisms that create herbicide resistance in crops. This is not surprising, since the chemicals used to kill weeds are functional analogs of the antibiotics used to control microbial infections in humans and animals, and the mechanisms of antibiotic resistance have long been a fruitful area of study for microbial physiologists.

The first clues for developing strategies for creating herbicide-resistant crops came from biochemical studies on weeds that had spontaneously mutated, in farmers' fields, into forms that did not respond to triazine herbicides. The high-level resistance of these plants was found to result from a single amino acid change in the herbicide receptor protein. This discovery involved a combination of several approaches, leading to the isolation of the protein from the membrane of the photosynthetic apparatus. X-ray crystallography then identified the exact structure of the site that bound the herbicide. This sophisticated information has aided in devising alternative weed control strategies for use where herbicide-resistant weeds have appeared.

In another case, the herbicide glyphosate was found to act on an enzyme required for the synthesis of several amino acids that are produced in plants but not in mammals, and the plant gene encoding the enzyme was cloned and then modified to replace an amino acid in the glyphosate binding site. A dramatic reduction in affinity for the herbicide resulted, without affecting the enzyme's activity in the plant.

This modified gene has now been inserted into a number of different crops to create glyphosate-resistant varieties.

A similar strategy was utilized to create sulfonylurea-resistant plants. In this case, however, a gene encoding the enzyme acetolactate synthase, the herbicide-binding site, was recovered from a herbicide-resistant weed. Transfer of this gene to crop species has produced herbicide resistance. In other studies, an enzyme in bacteria that degrades a herbicide has been identified and the corresponding gene has been transferred into crop plants. The bacterial enzyme destroys the applied herbicide without harming the plant.

The value of herbicide-resistant crops to the farmer will be primarily to add new options in weed control chemicals, as well as more effective control. For example, crops resistant to newer, degradable herbicides will accelerate the introduction of these chemicals, to replace older classes that persist in soil and are required in larger amounts.

Virus Resistance

Viruses are a persistent problem in crop production throughout the world. Because of the nature of their replication, no chemical control strategies have been developed to protect crops from viral diseases. Accordingly, genetic engineering to give plants virus tolerance or resistance is of major importance.

The basis of one form of genetic engineering for virus resistance is a phenomenon called cross-protection, in which infection of plants by a weakly virulent virus results in a type of immunity against a subsequent inoculation with a virulent virus. Moreover, this phenomenon has been simulated by inserting into crops a gene that causes the production in the plant cell of viral coat proteins, whose presence protects the plants against virulent virus strains. The mechanism for this resistance is not fully understood, but perhaps binding of the coat protein to the incoming viral nucleic acid inhibits its replication. Field tests of transgenic tomatoes carrying a gene for virus resistance have yielded promising results.

At least three different approaches to achieve virus resistance are now being tested. Some or all of these have a good chance of success for the major viral diseases. Creation of virus-resistant crops offers the promise of stable plant production in the many areas of the world where viral diseases have consistently reduced crop yields. This will be especially important in tropical countries, where these diseases often reach epidemic proportions.

Insect Resistance

For many plant species insects are the most important threat to crop production. With many vegetables even slight damage makes the product economically unacceptable. Boll weevil and bollworm infestations can devastate the flowers that yield cotton. In many agricultural areas chemical pesticides have been the main defense, but their potentially deleterious effects on the environment and on human health are of major public concern. As a result, extensive government regulatory activity has been directed at reducing the amount of chemical pesticides in the environment. To maintain a reliable and low-cost food supply, there is urgent need to find replacements for current methods of insect control. One approach in which progress has been achieved is engineering insect resistance in transgenic plants.

A strategy for creation of insect-resistant plants has been developed from an understanding of the life cycle of the bacterium *Bacillus thuringiensis*. When insects ingest the bacterial spore it germinates and produces a protein that blocks their digestive processes. The bacterium can then colonize the insect, cause its death, and subsequently enter sporulation to be released to start a new cycle.

Genes that encode insect control proteins have been cloned from *Bacillus thuringiensis*. When these genes are inserted into various plants, the insecticidal protein accumulates within plant leaves, stems, or roots. In field tests ordinary tomato plants inoculated with tomato fruit worm and pinworm were severely defoliated, while plants containing the *Bacillus thuringiensis* genes suffered no agronomic damage. Analogous field tests are now being conducted with other crop species. Particularly encouraging results have been observed in cotton.

Genetically engineered insect-tolerant plants will likely continue to involve the insect control proteins from *Bacillus thuringiensis*, but they may also incorporate genes for plant defense systems, such as plant proteins that block the insect's digestion. While it is unlikely that transgenic insect-resistant plants will soon totally supplant the need for chemical insecticides, they should make an important contribution to integrated pest management by lessening that need.

Biotechnology Regulation

Since the late 1970s many new biotechnology companies have been created to use genetic engineering for crop improvement. In addition, a number of large companies traditionally involved in agrochemical development have started their own biotechnology efforts. Government laboratories and universities are also committed to developing new tools for agriculture. Most of these efforts are focusing on the creation of plants with enhanced value due to unique production traits, such as resistance to herbicides, fungi, insects, or viruses. In addition, work on the "next wave" of biotechnology products is now expanding rapidly. This will create plants enhanced in such qualities as nutritional value and storage life. There is also a great deal of interest in using plants to produce specialty chemicals, such as flavors and industrial enzymes.

Many companies have made great progress in transforming basic science into products of potential value. The development of any new technology for commercial purposes, however, also creates the need to evaluate the products for negative attributes. Accordingly, guidelines have been developed, and are still being refined, for the field testing of transgenic crops.

The current guidelines have a number of controversial aspects. One is their emphasis on the process by which the gene transfer is conducted. Scientific investigation is frustrated by an excessive regulatory focus on the mechanism of transfer (recombinant DNA) and on the source of genes introduced (such as whether or not the new trait originated in a weed). This focus is unwarranted, since pathogenicity and the ability to colonize a natural environment are extremely complex processes that have been honed by evolution over long periods of time; they are not simple events to be "stumbled upon" by a molecular biologist. A regulatory focus on the product rather than the process of gene transfer would be more useful.

Current regulations of field tests also require extensive, and thus expensive, documentation. As a result, field testing is now almost restricted to large companies. More simplified procedures must be developed for small-scale testing, to allow academic researchers and smaller companies to participate in the "market preparation" stage of research for evaluation of potential products. In addition to small field tests, guidelines must be put in place for large-scale testing of transgenic plants, to allow their commercialization. Clear regulatory guidelines must also be established for confirming the safety of foods and food products produced from genetically modified plants.

Conclusion

There has been phenomenal progress in developing new protocols for plant genetic engineering in recent years. These tools are being used on crops to add traits that will cut production costs, stabilize productivity, and yield higher-value products. Many potential commercial products derived from cell culture or gene transfer technologies are already undergoing field tests. To ensure competitiveness in U.S. agriculture, the processes for regulating the introduction of genetically engineered crops must be continually improved, with the goal of realistically protecting the consumer and the environment while encouraging new technologies in the agricultural industry.

Epilogue: Comments on Chapter Five

Whereas I have viewed crops genetically modified in production traits, such as herbicide resistance, as a means to increase options for the farmer, the chapter by Margaret Mellon refers to this approach as an effort to "shackle [agriculture] to herbicide use for the foreseeable future." The following example of agricultural production practices may help readers to evaluate these different viewpoints.

All herbicides currently available to farmers began with an empirical evaluation of chemical selectivity on weeds versus crops. In practice this is accomplished by applying experimental chemicals to test samples of weed and crop seedlings. The compounds that show selective activity—causing death of weeds but not of one or more crops—are then subjected to animal toxicology studies. Those that are acceptable are developed for farmers' use.

The search for chemicals that show selective plant control is based on our knowledge that crops have mechanisms of resistance to various herbicides involving enzymes that convert the herbicides to inactive derivatives. These enzymes tend to be "crop-specific." For example, the enzymes that render soybeans immune to the action of thirty or more commercial herbicides are different from the enzymes that make corn resistant to other herbicides. This can be a benefit; when corn and soybeans are grown in rotation, the unwanted "volunteer" corn that germinates in a soybean field is controlled by the soybean herbicide.

The crop specificity in herbicide-resistance mechanisms can also create limitations. For instance, certain broadleaf weeds mimic soybeans in their complement of enzymes. With biotechnology, researchers are devising ways to give crops such as soybeans an additional enzyme to augment their natural herbicide-resistance mechanisms. This is not being done indiscriminately; the goal is to find ways to give the farmer a simpler means to control weeds through the use of a single, more effective herbicide selected on the basis of several traits, including reduced cost and greater safety in the environment.

This explanation may not satisfy a reader who is asking, "Why don't they simply stop using all herbicides?" The main answer is their huge benefit. Weed control is essential for all crops; if left unchecked, weed populations will reduce crop yields to near zero. Prior to the 1950s weeds were controlled strictly by mechanical means. By the 1980s, in contrast, more than 95 percent of all major row crops in the United States were produced using herbicides instead of manual cultivation. Farmers based this decision on costs: an application of herbicides at the time of planting, costing as little as $10 per acre, is many times cheaper than mechanical cultivation, which requires specialized equipment (capital costs), fuel for tractors, and intensive operator time (labor costs). The methodologies in place for weed control today are vastly cheaper per unit of production than they were in the 1940s (calculated on adjusted dollar values), and they are a factor in the low food prices we enjoy in the United States. They also save on energy utilization, and when properly used they do not harm the environment.

Since this epilogue includes an economic perspective, it is worthwhile to comment on Dr. Mellon's evaluation of technology related to other aspects of agriculture. She specifically criticizes the introduction of bovine somatotrophin (BST) because the United States and Europe suffer from a milk surplus, and this biotechnology product will allow the production of more milk per cow and per unit of feed intake. I do not agree that technology should be blamed for exacerbating a problem created by government subsidy policies. If that logic were valid, our current overproduction could be blamed on farmers' use of other technical innovations that have revolutionized dairy production, including automated milking machines and genetic improvement via artificial insemination. In fact, it is these innovations that have allowed farmers to provide such low-cost milk.

In the dairy industry, as in other phases of agriculture, producers strive to reduce costs, and BST has been shown to accomplish this. Moreover, in contrast to Dr. Mellon's statements, this technology is not farm-size-selective. Both large and small farms benefit by increasing herd output without increasing total fixed costs.

It is true, however, that BST, like other technologies, requires proper use for optimal dollar return. This includes recordkeeping to follow animal nutritional parameters to optimize BST dosage and timing. Some critics have argued that small farmers are not prepared for this level of sophistication, and therefore will be at a competitive disadvantage in technology adoption. I am not convinced that this is true. Even if it were, development of regulations banning new technology based on principles of protecting unskilled producers would be a dangerous precedent, in an era when the United States is coping with increasing global competition.

Farming is not simply a lifestyle; it is a major segment of our economy, with the unique feature of being dominated by small business units. It is not an activity at odds with nature or the environment; it is a set of activities that must maintain the natural resources on which it is based. (*Sustainable agriculture* is the term currently used to designate this concept.) And foremost, farming must be a business segment that is not static; it must develop and adapt new technologies to expand food and fiber production in a world with an escalating population crisis.

Agriculture of the 1990s will face a series of public perceptions that will have a profound impact on farming. An informed public will recognize that agriculture must meet global societal needs.

Additional Reading

Arntzen, C. J., and C. Ryan. *Molecular Strategies for Crop Protection*. New York: Alan R. Liss, 1987.

Borlaug, N. E. "Challenges for Global Food and Fiber Production." *K. Skags Lantbr. Akad. Tidskr.*, Suppl., 21 (1988): 15–55.

Borlaug, N. E., and C. R. Dowswell. "World Revolution in Agriculture." In *1988 Britannica Book of the Year*, pp. 5–14. Chicago: Encyclopaedia Britannica, 1988.

Kung, S.-D., and C. J. Arntzen. *Plant Biotechnology*. Stoneham, Mass.: Butterworth, 1989.

Mellon, J. W., and F. Z. Riely. "Expanding the Green Revolution." *Issues in Science and Technology* 6 (1989): 66–74.

Wittwer, S., Y. Youtai, S. Han, and W. Lianzheng. *Feeding a Billion*. East Lansing: Michigan State University Press, 1987.

8

Animal Agriculture

Franklin M. Loew

In 1887 the philosopher Jacob Gould Schurman wrote, "Man creates nothing; he waits for the variations which nature gives; and then selecting those which are useful or pleasing to him; he preserves them, and in preserving accumulates them . . . Man's power of accumulative selection is, therefore, the key to the origin of our diverse breeds of domesticated animals and cultivated plants."[1] And how the nineteenth-century agricultural scientists and gentleman-farmers did accumulate and select! Dozens of breeds of livestock and pets were painstakingly nurtured by careful, generation-after-generation selection of parents with desired traits: Hereford, Holstein-Friesian, Angus, and Shorthorn cattle; Yorkshire, Hampshire, and Berkshire swine; Suffolk and Shropshire sheep; Clydesdale, Thoroughbred, and Morgan horses; and, of course, King Charles spaniels, foxhounds, and Great Danes.[2] Well before the nineteenth century, Pekingese dogs, Siamese cats, Percheron and Arabian horses, and Jersey and Guernsey cattle had been purposefully developed.

Among the areas for application of the genetic revolution, domestication will be prominent, along with medicine and veterinary medicine, because animal products have great nutritional importance. In the United States they supply approximately two-thirds of the proteins, one-third of the energy, one-half of the fat, four-fifths of the calcium, two-thirds of the phosphorus, and significant essential minerals and micronutrients in the human diet. Conventional genetic selection, along with better nutrition and disease control, has already given us dramatic improvements in yields of products from livestock. Poultry broilers today can attain twice the weight on half the feed and in half the time of a broiler of twenty years ago, and milk production in dairy cattle has doubled per unit of feed over the last thirty years. Likewise, today's swine are leaner, more feed-efficient, and more fecund.

In a modern variant of domestication, breeding has also yielded animals with increased or decreased susceptibility to specific diseases, and their use as models has been vital to progress in medicine and veterinary medicine. An array of strains selected methodically for use in cancer, immunologic, or metabolic research is now available worldwide.[3]

Genetic engineering will now greatly accelerate the process of domestication. The ability to develop in one generation modifications that formerly required tens and hundreds of generations will be of enormous advantage both to agriculture and to science. But it is even more important that some specific traits can already be added to an animal's inheritance with great precision. The future will soon bring control of even more complex phenomena, such as specific structure of fats (e.g., saturated or unsaturated). But contrary to popular expectations, fundamental properties of a species, such as the number of legs, will not be subject to this sort of genetic manipulation, at least in the foreseeable future.

Transgenic Animals in Agriculture

A particularly powerful application of modern genetics to animal husbandry is the formation of transgenic animals. These are obtained by introducing a small block of a desired genetic material either into germ cells (which then form a fertilized ovum) or into an already fertilized ovum; the altered ovum is then implanted into a suitably prepared female animal. The genetic material may be inserted either mechanically, or by application of an intermittent electrical field, or by infection with a virus carrying the genetic material; the last is called transfection. When the introduced DNA is integrated into a chromosome, the animal and its offspring carry the new gene. So far all of these processes are inefficient, both because only a small proportion of the treated germ cells incorporate the new gene and because it often yields little of its product. However, both these problems are likely to be solved soon.

Unlike conventional breeding methods, transgenism is not limited to genes from the same species. Greatly broadening domestication, genes for a certain trait from other species, including nonfarm animals, can be transferred to farm animals.

One of the first genes used to make transgenic animals was the gene for growth hormone, also called somatotrophin or somatotropin. Palmiter and co-workers found that when germ cells from mice incorporated this gene from a rat, along with regulatory DNA that caused great overproduction of the hormone, the resulting animals grew faster and became much larger than their nontransgenic littermates.[4] Similar results were obtained later with human growth hormone gene. These dramatic products, called "mighty mice," prompted a flurry of activity in research on other species.

In agricultural animals transgenism has been used to bring about changes in meat and in milk, but the effects cannot yet be precisely controlled. Pigs with a gene that provides extra growth hormone grow larger and have more muscle (meat) and less fat; but the results have been variable, and there are often serious side effects, including skeletal, muscular, and metabolic abnormalities and impaired fertility.[5]

More complicated traits are even less likely to be changed easily, but such changes are probable within the coming years. For example, reduction in the saturation of fats would bring beef and pork into better conformance with what is now recommended for human diet. Similarly, development of animals with increased resistance to infections not only would reduce disease but also would reduce or eliminate the present widespread use of antibiotics in livestock feed to promote growth, probably by mediating changes in the population of gastrointestinal microbes that may involve mild pathogens. Although there is controversy over whether this practice leads to antibiotic-resistant bacteria that spread to humans, and to harm from residues of antibiotic in the meat or milk, there is no controversy over the money farmers would save if they no longer had to buy antibiotics.

In contrast to such effects, brought about by one gene or very few, genetic engineering is unable to modify enough genes to alter the fundamental nature of an animal. Five-legged horses or two-headed cattle are simply not possible using genetic engineering. Monstrosities affecting basic anatomy have always occurred in nature, but they result from poorly understood environmental effects on development of the embryo, rather than from genetic changes.[6]

Because most agriculturally important traits are controlled by multiple genes, they are likely to continue to be the subject of classical selection, which randomly reassorts these genes into novel combinations. Mules, which are hybrids of female horses (*Equus caballus*) and male donkeys (*Equus asinus*), have been produced for centuries, and they represent a much larger alteration of species than can be expected from simple gene insertions.[7] Intermediate between mules and single-gene transgenic animals are chimeras, produced not by fusing germ cells of two species but by mixing cells of two or more compatible species in an embryo. For example, a goat-lamb resulted from combining blasto-meres (early embryos of only a few cells) from a sheep and from a goat.[8] This removal of the reproductive barrier between sheep and goats repre-sents a novel procedure in reproductive biology, rather than in genetic engineering; but such manipulation of cells provides powerful tools for applying molecular genetics to developmental research.

In newer areas of agriculture, such as fish farming (aquaculture), ge-netic alterations are already leading to improvements. Because fish eggs develop outside the mother's body, new genes can be injected into them easily, and no recipient female is required for their development. Carp and catfish genetically engineered for various desirable traits will prob-ably be in commercial production by 1994.[9] The Food and Agriculture Organization of the United Nations estimates that by the year 2000 fish will be taken from the oceans faster than they can replenish them-selves.[10] Under such circumstances the farming of fish will become a necessity lest wild species be depleted. In the United States, never a major fish-eating country, per capita fish consumption has increased from 13 pounds in 1979 to 15.4 pounds in 1987, while beef consumption has decreased from 78 pounds to 73.4 pounds.

In another development, animals can be used as "biological facto-ries" for the manufacture of specific desired proteins, which can then be more easily recovered than from bacteria. This approach became possi-ble with the identification of a regulatory DNA sequence that becomes active only in mammary tissue, where it "turns on" the regulated gene, whose product is then secreted in the milk. In the first such experi-ments, in mice, the regulatory sequence was artificially attached to a gene for a protein, human tissue plasminogen activator, that accelerates the dissolution of blood clots and hence is useful in treating heart at-tacks. These transgenic animals secreted large amounts of the protein into their milk.[11]

Existing technology will allow this method to be extended from mice to species that produce large quantities of milk, such as cattle, sheep, and goats. Indeed, genes have been introduced into sheep to yield products in their milk. Genetic engineering of these species, called molecular farming, can result in the production of large quantities of scarce, exotic proteins[12]—at present proteins useful in fighting disease, but no doubt many more in the future.

Transgenes are also being studied to try to cause desired chemical changes in chicken eggs, such as reduction in cholesterol content. Con-sumer concerns over dietary cholesterol led to a decreased consump-tion of eggs in the United States from 288 per capita in 1973 to 249 in 1987. Genetic engineering also might be used to modify the composition of milk so that it can be consumed by patients with various dietary restrictions.

Animals in Research

Transgenic animals promise to be invaluable as subjects in medical research. In addition, before virtually any products of research can be used in medicine, they must first be tested in animals for safety and for efficacy, and this will be true of products of genetic engineering in animals. It therefore seems useful to provide the perspective of a veterinarian on animal research, which has become one of the most contentious issues facing science and society.

The first federal law dealing with animals in research, the Laboratory Animal Welfare Act, went into effect in 1967. This law controlled only dealings in, and acquisition of, laboratory dogs and cats. Whereas earlier initiatives had failed, the act was passed, after stormy hearings in both the House of Representatives and the Senate, largely because of a public scandal about animal dealers who kept their dogs and cats under deplorable conditions. The act was very limited in its intent.

Since then public concern about the use of animals has steadily increased,[13] and several amendments have been added to broaden the scope of the original law. These changes address the care of animals in research facilities, including increased accountability to the National Institutes of Health (NIH) and other federal agencies that support research.

This same period has also generated regulations over other aspects of research, including the use of human subjects and work with recombinant DNA. The "animal issue," however, is different in several ways. For two hundred years a small but impassioned and vocal segment of the public has been protesting the use of "helpless and innocent" animals, especially of species used as pets, in laboratories. These voices were represented in England in the nineteenth century by the antivivisection societies, and the movement was supported in the United States by William Randolph Hearst's newspapers in the first half of this century. In recent years, however, a much broader segment of the public, including some philosophers and scientists, has begun to voice concern about animal research. Moreover, like the abortion issue, the controversy evokes a depth of passion, from both protesters and supporters, that is not seen in the debates over other areas of regulation.

Some critics are interested in animal welfare, aiming at more humane treatment of animals in research as well as elsewhere. Others, however, under the banner of animal rights, are opposed to all use of animals to serve human needs, however humanely carried out; and they would even prohibit the use of mice. This group has found that the research community, with its relatively few votes, is a more favorable target than agriculture, though the scale of agricultural use is much larger.

About 250,000 dogs and cats are used in research annually. In contrast, 10–13 million unwanted cats and dogs are destroyed in the nation's pounds and animal shelters annually. Moreover, some 5 billion chickens are consumed as food annually in the United States, and there are about 90 million cattle and 60 million swine on the nation's farms.[14]

In fact, of the 18–22 million animals estimated to be used annually in medical, biological, and agricultural research, teaching, and testing, more than 85 percent are mice and rats bred specifically for the purpose.[15] Mice have the particular advantage that they can be used to mimic various specific genetic defects that cause disease in humans.[16] These animal "models" are useful both in studying the diseases and in screening possible treatments. Morever, the principles used to induce genetic diseases in mice are a model for genetic changes with the opposite effect: gene therapy for human genetic diseases.

A few research laboratories have been found to care poorly for experimental animals, and their exposure has drawn public attention and has put the scientific community on the defensive. The media and members of Congress are taking seriously charges that scientists are mistreating their experimental animals.[17]

In response to the new sensitivities, scientists have generally admitted that the Animal Welfare Act needs improvement. In 1985 both the act and the NIH's requirements for its grantees were made more stringent.[18] Replacing the relatively broad freedom that scientists used to have in deciding what they needed to do to animals in their research, research institutions must review and approve all animal projects prior to their start. In addition, much greater attention is being paid to the reduction of animal pain and distress, and there is an informal but growing tendency to attempt to weigh the costs of the research, in terms of animal distress and death, against the anticipated benefits, for medical and veterinary practice and for knowledge.

There is still considerable confusion over the extremely stringent regulations that are being proposed under the latest amendments. Of particular concern, for example, are a requirement that dogs must be exercised daily despite a virtual absence of data showing such a need, and a vague requirement that the "psychological well-being of nonhuman primates" must be provided for. Moreover, institutional committees charged with animal care and use are finding it difficult to answer such seemingly simple but in reality complex questions as evaluating either the extent of animal pain and distress or the potential benefits of the research. Like nearly all regulations, these pose costs to society. Science groups estimate that the recent proposed amendments to the act might cost as much as $1 billion per year. Whether this was the original intent of animal activists or not, the increased standards are seriously raising the dollar costs of much agricultural, pharmaceutical, medical, and other life sciences research.

While the levels of regulatory control over animal research in the various developed nations may be similar, differences in traditions still may lead to somewhat different approaches. For example, in the United Kingdom the individual scientist has to have a license to perform animal research, while in the United States it is the institution that is responsible. Britain certainly compares well with the United States in Nobel Prizes in physiology or medicine, although some activities, such as surgical innovation and trauma investigation, are more constrained there. Moreover, a new Animals (Scientific Procedures) Act, passed in 1986, will require that not only each scientist but each project be licensed. In addition, all projects must be subjected to a cost-benefit analysis.

Though scientists have generally supported legislation designed to ensure improved animal care in laboratories,[19] the abolition of animal use altogether is a different matter. The latter attack arises from the notion that animals have rights, such as the right not to be eaten or to be experimented upon.

"Animal rights" is not merely the extreme end of the traditional animal welfare spectrum, but an entirely new way of looking at animals. For example, Peter Singer, an Australian philosopher, has challenged the traditional dominion of human beings over animals.[20] The notion that animals have "rights" in the same sense as people has been surprisingly popular, especially among educated Americans. That its premises are essentially unprovable (see, e.g., Cohen or Fox)[21] has not deterred its adherents. Singer has emphasized that there is no logical basis for the assertion that we have the right to use animals. This is true; but there is also no logical basis for his assertion that we have no such right. We are dealing with value judgments, and society can settle for whatever standards it wishes at a given time. But while the advance of civilization promotes an admirable trend to decrease cruelty, a realistic solution must balance its definition of cruelty with the instincts and needs derived from our evolution as a carnivorous species.

Unfortunately, the noble if extreme ideal of animal rights seems to meet an emotional need of people who seek something that transcends their material satisfactions—a sort of secular equivalent to religious fundamentalism. And like the extremes of fundamentalism, it has led to some violent attacks on the objects of its hate. The researchers who oppose this movement are made to seem self-serving; and the main group of beneficiaries—the whole population that has benefited from modern medicine—has so far provided very little organized opposition.

The Patenting of Transgenic Animals

Particular objections have been raised to the patenting of transgenic animals. When the Office of Technology Assessment (OTA) analyzed this issue in late 1988, there were nine pending patent applications; four months later there were sixty-five.[22] The OTA identified four categories of objection to patenting: metaphysical and theological issues; inappropriate treatment of animals; damage to the environment; and economic burden on agriculture. Virtually all these objections are informed by preconceived values and beliefs, with little incorporation of scientific or economic facts. The OTA, in fact, concluded that "it is unclear that patenting *per se* would substantially redirect the way society uses or relates to animals." The patented "Harvard mouse," genetically engineered for great susceptibility to a particular cancer, has already found a role in cancer research.

The objection concerned with the environment assumes that patenting will somehow promote cultivation of animals that may spread and "take over." But since transgenism is an extension of the ancient practice of domestication, there is no evident reason to fear the development of monsters. In domestication, the animals with desirable characteristics are retained; the others are discarded; and people control the distribution of the new strains.

In the economic objection, farmers and some of their organizations have expressed fear that patenting will result in a prohibitively high cost of animals. However, unless a net economic benefit reaches the livestock producer, such animals will not find a market. It is therefore not obvious how patents will place farmers at a disadvantage. The related issue of whether the use of patented animals requires the payment of a licensing fee has not yet been settled, but it surely can be.

There are ethical issues related to genetic modifications of animals. They were a subject of concern for Schurman in 1887, and they are even more important now. Scientists must address them satisfactorily. The most substantive has to do with any unanticipated side effects of transgenism that lead to discomfort. But as we have already seen, the gene for overproduction of growth hormone in pigs had painful side effects, and these animals are not being patented or developed for commerce.

Bacteria and plants have long been patented without controversy. In a similar mechanism of control, for many years breeds of horses, cattle, and dogs have been registered, for a fee, by private organizations: the

Jockey Club, Holstein-Friesian Association, American Kennel Club; only registered animals can be raced or entered in shows in the United States. Breeders can obtain registered stock only by buying, at market prices, from other breeders or owners of duly registered stock. An off-spring of two registered Thoroughbred horses cannot race on any U.S. racetrack, nor can any of its offspring, until registered by the Jockey Club. With such precedent, it is hard to see the logic or the benefit of treating transgenic animals as an exception.

Other Applications of the New Genetics to Livestock

Genetic engineering is useful for animal husbandry not only through gene transfer but also through other developments: use of engi-neered products in animals; cloning of superior animals by nuclear transfer in embryos; and production of materials useful for vaccination against or diagnosis of infectious diseases of animals.

Hormones produced by genetically engineered microorganisms promise more immediate use in farm animals than gene transfer, and they present fewer difficulties. Bovine and porcine growth hormones (somatotrophins) are used to increase milk yield in cows and growth in pigs, respectively. Although these effects have been known for a long time, only recently has their manufacture by bioengineering made it economically feasible to use them in agriculture. Experimental injec-tion of bovine growth hormone into dairy cows has resulted in increases in feed efficiency, and increases in milk yield up to 40 percent have been observed, though under farm conditions 8–20 percent is more realistic. There is thus a strong economic incentive to adopt this new technology.

Despite the benefits of the use of bovine growth hormone to both farmers and consumers,[23] some farmers' organizations and agricultural officials from some states are attempting to stop its use. They argue that since there is already a milk surplus in the United States, this hormone is an unnecessary technological aid that will hasten the demise of America's small farms. This argument ignores the fifty-year decline in the number of farms, as less efficient farmers—like less efficient busi-nesspeople everywhere—enter other occupations; and it also ignores the desperate need for food in many parts of the world. In fact, bovine growth hormone will lower farmers' production costs and hence con-sumer prices, as is recognized in the larger farm states; and since small farms can use the product as easily as large ones, it poses no particular threat to them.[24]

Porcine growth hormone can increase the rate of weight gain in pigs 10–20 percent and their feed efficiency 15–30 percent, while reducing their backfat by 70 percent.[25] The bovine hormone also stimulates growth in beef cattle. However, while the required daily injections are feasible in dairy farming, they cannot be carried out economically in present-day beef and hog management systems. Slow-release dosage forms, requiring only infrequent injections, are expected to be available soon.

Another recent development with great promise is cloning by nuclear transfer.[26] An early embryo is dissociated into its individual cells, and the nucleus from each is transferred to replace the nucleus of an egg cell previously recovered from a donor animal. These cells are then each placed in the reproductive tract of a recipient animal to develop. The resulting offspring are genetically identical; that is, they constitute a clone. Its size is limited because the initiating embryo can go through only a few cell divisions before the cell nuclei lose the capacity to initiate a new embryo.

This kind of cloning copies a fertilized egg, and so genes of the two parents have been randomly recombined. As with human identical twins, derived from the same egg, the products are unpredictable. In contrast, cloning from the cells of an adult, if feasible, would yield progeny with a performance record and a predictable set of genes; but this possibility is extremely remote. Indeed, evidence for changes in the genome of differentiating cells suggests that it is probably impossible.

Cloning from embryos has been extended from laboratory animals to cattle.[27] There is great interest in using this technique to amplify strains of special agricultural value—not only transgenic animals with a useful foreign gene but also the offspring of outstanding sires or dams. It is likely that the process will be perfected and used commercially within a few years.

There are concerns that widespread use of such a method could reduce genetic diversity. However, the reservoir of diverse genes need not be lost, since it is now possible to preserve genetic material in the form of frozen germ cells or embryos.

A third major area of application of genetic engineering to animal agriculture is the development of effective new vaccines and reliable diagnostic tests to reduce the prevalence of infectious disease, which results in losses of about $14 billion annually in the United States. One improvement is the development of subunit vaccines, which use a single chemical component from the infective virus or bacterium instead of the whole organism in its traditional killed (inactivated) form or in a live, attenuated form. To produce such vaccines the gene for a specific immunogenic antigen, such as a surface protein or a particularly effective part of that protein, may be inserted into a bacterium or yeast for manufacture of the antigenic substance in large quantities. Alternatively, small portions of a protein can be chemically synthesized. Either kind of material stimulates antibody formation when injected into an animal. Subunit vaccines have advantages over live, attenuated virus: they cannot cause accidental infection, and they may be more stable and may elicit a stronger immune response.

Alternatively, live pathogens can be rendered safe for use as attenuated vaccines by removing known genes essential for producing disease: a better procedure than empirical selection. This method has already been used to produce a bioengineered vaccine for pseudorabies, an important disease of swine.

Vaccinia virus, a variant of cowpox (Latin *vacca*, cow), is now used as a delivery system for other vaccines. This relatively large, nonpathogenic virus has been used for two centuries as a live vaccine (hence the origin of the word *vaccine*) to convey a strong, long-lasting immunity to the closely related smallpox virus, and it is easily and cheaply manufactured and administered. When genes for proteins from other pathogenic viruses, or from bacteria, are incorporated into vaccinia, it can be used as a vaccine that codes for those proteins in the transiently infected cells. Because vaccinia is such a large virus, it is possible to insert DNA from several pathogens into it and thus immunize animals against more than one disease at a time.

These new kinds of vaccines may eliminate the time-honored but expensive animal disease control strategy of "test and slaughter." For several important indigenous diseases (such as tuberculosis, brucellosis, and bluetongue) and imported diseases (such as foot-and-mouth disease and African swine fever), a positive diagnosis has mandated immediate isolation or slaughter to prevent spread of the disease and the consequent depression of productivity, or the loss of export markets due to fear of importation of the pathogen.

The practical use of vaccines in livestock will depend on the availability of diagnostic tests that can differentiate between the antibodies in vaccinated animals and those in infected animals. Unfortunately, earlier, whole-organism vaccines elicited the same set of antibodies as the disease; hence, even though vaccination would often be a more cost-effective as well as humane strategy than slaughter, it could not be used. However, with the subunit vaccines the animal forms antibodies only to the administered antigen, and not to the many other antibodies that can be detected in the diseased animal. Morever, genetic engineering can be expected to improve these diagnostic reagents. Hence, vaccination of agricultural animals is likely to become widespread in the near future.

Conclusion

The history of agriculture has consistently been one of selective breeding for domestication. In this context, the ability to speed up selective breeding by gene manipulation is not a qualitative departure from traditional animal husbandry. It promises to lead to more value-added milk, eggs, and meat; to more efficient production; to more reliable and humane disease prevention, diagnosis, and control in livestock; to valuable models for research on human physiology and disease; and to use of animals as "factories" for the manufacture of valuable proteins even more cheaply than in bacteria or yeasts. And just as the traditional biotechnology of domestication has been remarkably free of untoward side effects, compared with the toxic products of some of the physical technologies, there is no reason to expect the animals modified by genetic engineering either to be a menace or to displace natural fauna.

Objections to the production or the patenting of transgenic animals are based largely on metaphysical convictions or on purely hypothetical dangers, and they have so far not been persuasive to most Americans or their elected representatives. This powerful new approach to improving livestock has produced transgenic fish, mice, rabbits, swine, sheep, goats, and cattle, and more will follow. A more serious impediment to the benefits from genetic engineering is the growth of an animal welfare movement that presses for excessively stringent regulation of research with animals, and an animal rights movement that objects to all use of animals to serve human society—now in research, and tomorrow no doubt in agriculture.

Humanitarian issues, in the usual sense of that word, do not necessarily arise from transgenism, cloning, or other biogenetic techniques unless these result in discomfort. In fact, swine injected with growth hormone had problems in joints, presumably associated with growth rates that outstripped the legs' ability to bear the weight; and these transgenic animals are not being commercially developed.[28] Thus, with the new approach, as in classical animal breeding, we can expect novel strains to be introduced into agriculture only when their overall set of characteristics makes it worthwhile. There is no reason to expect that the result will be cruel, any more than the development of the already familiar varieties of domesticated animals.

Notes

I thank Donald L. Black of the University of Massachusetts for assistance with some aspects of this chapter.

1. J. G. Schurman, *The Ethical Import of Darwinism* (New York, 1887), p. 55.

2. H. Ritvo, *The Animal Estate* (Cambridge, Mass.: Harvard University Press, 1987).

3. Among domestic animals, members of the same species (e.g., the dog, *Canis familiaris*) differing in appearance of some properties are called breeds (e.g., cocker spaniel, German shepherd), while the analogous term for laboratory rodents (e.g., the mouse, *Mus musculus*) is strain (e.g., C_3H, C_{57}). An inbred strain is one that results from at least twenty generations of parent-offspring or sibling-sibling matings, resulting in homozygosity for most genes.

4. R. D. Palmiter, R. L. Brinster, R. E. Hammer, M. E. Trumbauer, M. G. Rosenfeld, N. C. Birnberg, and R. M. Evans, "Dramatic Growth of Mice that Develop from Eggs Microinjected with Metallothionein–Growth Hormone Fusion Genes," *Nature* 300 (1982): 611–15.

5. V. G. Pursel, C. E. Rexroad, Jr., D. J. Bolt, K. F. Miller, R. J. Wall, R. E. Hammer, C. A. Pinkert, R. D. Palmiter, and R. L. Brinster, "Progress on Gene Transfer in Farm Animals," *Vet. Immunol. Immunopathol.* 17 (1987): 303–12.

6. F. B. Hutt, *Animal Genetics* (New York: Ronald Press, 1964), p. 473.

7. R. Pong, H. Cai, X. Yang, and J. Wei, "Fertile Mule in China and Her Unusual Foal," *J. R. Soc. Med.* 78 (1985): 821–25.

8. S. Meinecke-Tillman and B. Meinecke, "Experimental Chimaeras: Removal of Reproductive Barrier between Sheep and Goat," *Nature* 307 (1984): 637–38.

9. Estimate of the Food and Agriculture Organization of the United Nations.

10. See, for example, *Chronicle of Higher Education*, 10 May 1989, p. A6.

11. K. Gordon, C. Lee, J. Vitale, A. Smith, H. Westphal, and L. Henninghausen, "Production of Human Tissue Plasminogen Activator in Transgenic Mouse Milk," *Bio/Technology* 5 (1987): 1183–87.

12. J. Van Brunt, "Molecular Farming: Transgenic Animals as Bioreactors," *Bio/Technology* 6 (1988): 1149–54.

13. Antivivisectionism is not new. See, for example, D. French, *Antivivisectionism and Medical Science in Victorian Society* (Princeton: Princeton University Press, 1975), for its European roots and history, and J. Turner, *Reckoning with the Beast* (Baltimore: Johns Hopkins University Press, 1980), for the U.S. counterpart. (My own review may save the reader time: F. M. Loew, "Animal Experimentation," *Bull. Hist. Med.* 56 [1982]: 123–26.) On animal rights, see P. Singer, *Animal Liberation* (New York: N. Y. Review of Books Press, 1975); M. A. Fox, *The Case for Animal Experimentation* (Berkeley: University of California Press, 1986); C. Cohen, "The Case for the Use of Animals in Biomedical Research," *N. Engl. J. Med.* 315 (1986): 865–70.

14. U.S. Department of Agriculture, *Agricultural Statistics* (Washington, D.C.: U.S. Government Printing Office, 1988).

15. Office of Technology Assessment, *Alternatives to Animal Use in Research, Testing and Education*, OTA-BA-274 (Washington, D.C.: U.S. Government Printing Office, 1986). According to the National Institutes of Health, 43.2 percent of its 1988 extramural research programs used mammals.

16. H. Westphal, "Transgenic Mammals and Biotechnology," *FASEB J.* 3 (1989): 117–20.

17. S. R. Kellert, "American Attitudes toward and Knowledge of Animals: An Update," *Int. J. Stud. Anim. Prob.* 1 (1980): 87–119; S. R. Kellert, "Human-Animal Interactions: A Review of American Attitudes to Wild and Domestic Animals in the Twentieth Century," in *Animals and People Sharing the World*, ed. A. N. Rowan (Hanover, N.H.: University Press of New England, 1988).

18. See the hearings records for amendments to the Animal Welfare Act: House Subcommittee on Science, Research, and Technology, 13–14 October 1981; House Subcommittee on Health and the Environment, 9 December 1982; House Subcommittee on Department Operations, Research, and Foreign Agriculture, 19 September 1984.

19. Ibid.

20. Singer, *Animal Liberation*.

21. Cohen, "Case for the Use of Animals"; Fox, *Case for Animal Experimentation*.

22. Office of Technology Assessment, *New Developments in Biotechnology: Patenting Life—Special Report*, OTA-BA-370 (Washington, D.C.: U.S. Government Printing Office, 1989). See also "Patent Applications for Animals Up Sharply," *New York Times*, 22 April 1989.

23. Extension Service, USDA, Session 4, Economic and Social Impacts, National Invitational Workshop on Bovine Somatotrophin, St. Louis, Mo., 1987. See also *New York Times*, 5 May 1989.

24. Ibid.

25. T. D. Etherton, J. P. Wiggins, D. M. Evock, D. S. Chung, J. F. Rebhun, P. E. Walton, and N. C. Steele, "Stimulation of Pig Growth Performance by Porcine Growth Hormone: Determination of the Dose-Response Relationship," *J. Anim. Sci.* 64 (1987): 433–43.

26. J. M. Robl, R. Prather, F. Barnes, W. Eyestone, D. Northey, B. Gilligan, and N. L. First, "Nuclear Transplantation in Bovine Embryos," *J. Anim. Sci.* 67 (1987): 642–47.

27. J. M. Robl and S. L. Stice, "Prospects for the Commercial Cloning of Animals by Nuclear Transplantation," *Theriogenology* 31 (1989): 75–84.

28. V. G. Pursel, C. A. Pinkert, K. F. Miller, D. J. Bolt, R. G. Campbell, R. D. Palmiter, R. L. Brinster, and R. E. Hammer, "Genetic Engineering of Livestock," *Science* 244 (1989): 1281–88.

9

Molecular Medicine

Theodore Friedmann

The past century has seen explosive growth in the understanding of biological phenomena, and the process is now greatly accelerated by the concepts and tools emerging in molecular genetics. We can confidently expect these developments to lead to extraordinary advances in our insights into the biological nature of human beings. It is no hyperbole to say that the implications will be as profound as any in the history of human endeavor.

This progress has been exciting and challenging but also troubling, since it promises, or threatens, to overturn many of our views of ourselves and our relation to the universe. For we are on the verge of two staggering achievements: the complete molecular description of our genetic heritage,[1] and the capacity to bring about designed changes in human genes. From the perspective of medical science, whose aim is to apply the most promising discoveries as quickly as possible to the relief of suffering from disease, these two sets of genetic advances—the Human Genome Project and the potential for gene therapy—will have important consequences.

Human Genetic Disease

In 1903 Sir Archibald Garrod provided the first scientific understanding of the relationship between genes and human disease in his description of "inborn errors of metabolism" such as albinism, a defect in the development of skin coloration and in eye function, and cystinuria, a metabolic defect that leads to the development of kidney stones. Garrod suggested that such hereditary metabolic defects resulted from the presence of errors in the genetic factors responsible for the production of "ferments," or enzymes as we know them now. In 1949, Linus Pauling and his colleagues discovered the mutation in the protein molecule hemoglobin responsible for sickle cell anemia. This was the first link of a defective gene to an aberrant protein—the first description of a "molecular" disease. In the 1950s and 1960s a large number of additional human diseases came to be associated with changes in specific genes and their protein products. By 1958, 412 genetic disorders were recognized, and in 103 of them an affected protein enzyme was identified (see Figure 9.1). The number has since grown to a total of almost four thousand recognized genetic disorders and more than two hundred identified enzymic defects.[2] Defects have also been demonstrated in many other, nonenzymic protein gene products, such as hemoglobin, factors responsible for blood coagulation and other circulating proteins, the collagen component of connective tissues, hormones, and many other functions.[3]

In addition to these diseases, which can be understood by simple and classical Mendelian inheritance as deficiencies of single genes, many and possibly most other diseases involve multiple genes interacting with each other and with environmental factors. These disorders include some of the most prevalent and burdensome human diseases, such as most cancers and many forms of neuropsychiatric, degenerative, and cardiovascular disease. The task of identifying the combined and interactive biochemical and environmental factors in complex multigenic diseases might seem daunting. But while the presence of more than one responsible gene in these disorders makes their characterization more difficult, work with multigenic traits in other organisms suggests that a small handful of different genes may play a major role in a complicated disease trait such as hypertension or diabetes.

While the detection of an abnormal enzyme or an aberrant metabolite is valuable for diagnosing and understanding some hereditary diseases, many others have remained beyond the reach of this approach. In many cases there are few or no clues to the nature of the abnormal metabolic step, as in Huntington's disease and cystic fibrosis. Without an assay for the defective enzyme or metabolite, the development of accu-

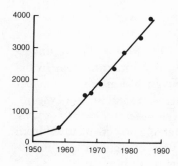

Figure 9.1   Number of recognized human genetic disorders, beginning in 1950. (Data from V. A. McKusick, *Mendelian Inheritance in Man* [Baltimore: Johns Hopkins University Press, 1988].)

rate and reliable diagnosis, prevention, and therapy has been difficult or impossible. For some disorders the expression of mutant genes may lead to irreversible damage very early in life, and the need to await inevitable physiological or cellular damage and the establishment of overt clinical disease before diagnosis or genetic counseling may destine patients to an unfavorable outcome. For instance, children born with Tay-Sachs disease seem perfectly well at birth, but they begin to show gradual and unrelenting neurological deterioration early during their first year of life and invariably die within several years.

Unfortunately, even with the availability of techniques for the detection of genetic disease, truly effective therapy is rarely achieved. For a few genetic diseases there have been therapeutic successes: the partially effective dietary therapies for phenylketonuria and galactosemia, both of which would cause severe mental retardation without prompt and early treatment; vitamin therapy for methylmalonic acidemia and multiple carboxylase deficiency, severe and lethal disorders of the newborn period; gene product replacement for forms of insulin-dependent diabetes mellitus, the bleeding defects in hemophilia, and the cretinism and mental retardation resulting from thyroid hormone deficiency; drug therapy for removing toxic accumulations of copper in tissues of patients with Wilson's disease; and even tissue transplantation for forms of thalassemia and for a variety of diseases caused by an accumulation of excess metabolic products in the brain and other tissues, such as in Hurler's syndrome (gargoylism). But most genetic diseases cannot be well treated by manipulation of the relevant metabolic pathways.

Diagnosis by Mutant Genes

The development of the tools of molecular genetics has led to great progress in identifying and purifying the genes responsible for diseases and in understanding their causative role in human genetic diseases. Until recently the isolation of specific genes generally required prior characterization of the enzyme or other protein product of the gene and some knowledge of its biochemical properties. This traditional approach has come to be called "forward" genetics, in which one must know something about the abnormal gene product before one can isolate the gene. Forward genetic approaches require:

1. Knowledge that a disease involves a genetic factor, most often established by family studies.
2. Identification of the biochemical or metabolic defect.
3. Purification, and determination of at least partial amino acid sequence, of the mutant protein.
4. Use of that sequence to prepare materials to help isolate the responsible gene from cloned gene "libraries" and then make and purify large amounts of the responsible gene ("clone" the gene).
5. Use of portions of the gene sequence to prepare materials for diagnosis and possibly for therapy.

For a growing number of human disease-related genes this forward, protein-based approach has already proved to be powerful and useful. The DNA "probes" isolated by these procedures can be used not only to diagnose disease in individual patients but also to screen large selected target populations for common and major diseases, such as sickle cell anemia in blacks and the thalassemias in Mediterranean populations. However, different defects within a single gene can cause the same clinical disorder, and when a disease has such genetic heterogeneity it is difficult to design diagnostic probes useful for large populations.

In addition to providing more effective diagnosis, the availability of cloned genes has made it possible to produce enzymes, hormones, and other products in virtually unlimited supply for use as therapeutic materials. The genes have been introduced into bacteria for large-scale preparation of medically important proteins, such as the antihemophilia factors, human growth hormone, human insulin, and tissue plasminogen activator (TPA) for the prevention of damage from heart attacks. Cloned genes can be similarly used for many other therapeutic reagents.

The Human Genetic Map

Unfortunately, the vast majority of human genetic disorders have no identified biochemical defect and no known target gene products, so they are not amenable to the forward genetic approach. To identify and characterize genes associated with such diseases, the combined powers of recombinant DNA technology and DNA sequencing produced an approach called "reverse" genetics, which is conceptually the reverse of the traditional forward scheme. It uses as its starting point the isolation

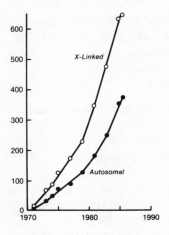

Figure 9.2 Number of genetic loci whose chromosomal locations have been established, for both the X chromosome and all the other chromosomes, called the autosomes. The sharp increase in 1979–80 corresponds approximately to the advent of restriction fragment length polymorphism linkage analysis. (Data from McKusick, *Mendelian Inheritance in Man.*)

of pieces of the genome sequences near a gene of interest, followed by various procedures to extend the sequence determination from that point until one reaches the actual gene of interest. The process is called genome "walking." Since disease can result from mutations not only in the genes themselves but also in sequences that regulate their expression, the end point of such a "walk" might be either an ordinary gene or a regulatory piece of DNA whose functions must then be determined.

The reverse genetic approach at first seemed extremely cumbersome, but it has now become one of the most powerful, effective, and promising approaches to understanding the bulk of human genetic disease. Its success depends not only on efficient cloning and chromosomal walking methods, but crucially on the development of an extensive description—a map—of the precise location and arrangement of genes and regulatory sequences along the human chromosomes. Very elegant methods have been developed to determine such a map by measuring the frequency with which genes are linked during egg or sperm formation (see Chapter 2). However, human gene mapping was virtually nonexistent even as recently as 1971, by which time only fifteen human genes had been localized to specific chromosomes (see Figure 9.2), twelve of them to the X chromosome (more easily identified because it is present in only one copy in males). From 1971 to 1986 there was a rapid increase in the chromosomal assignments of genetic loci. By 1979 there were 230 confirmed genetic markers on the X chromosome and 123 on the other 22 pairs of chromosomes, the autosomes.

In 1978 it became evident that the human genome contains markers other than traditional gene loci that can be used for the determination of a genetic map of a complex genome. These genetic markers are the normal minor variations (called polymorphisms) in nucleotide sequence that are found scattered at hundreds of thousands of positions throughout the genome. Many of these changes are not deleterious to any vital function and are therefore not associated with any defect or disease. Some occur with the short sequences that are the sites of cleavage by restriction endonucleases. These are the enzymes that have played a pivotal role in the development of recombinant DNA technology, because they make reproducible and defined cuts in DNA that allow the splicing and reconnection of segments of DNA. Because individuals differ in the distribution of the cleavage sites in many regions of DNA, the sizes of the DNA fragments produced by digestion of the genome DNA with the same restriction enzyme differ among individuals. In other words, there are restriction fragment length polymorphisms (RFLPs), and RFLPs can be used as markers for genetic mapping just like any other genetic loci.

Because the defective and the normal gene, on a homologous pair of chromosomes of a carrier parent, are often located on different RFLPs, analysis using RFLPs or other markers in a sufficiently detailed human genetic map provides the opportunity to identify the presence or absence, as well as the map location, of a disease locus, even in instances where the nature of the biochemical defect is totally unknown. Striking successes have occurred recently in diagnosing such previously incomprehensible diseases as Huntington's disease, Duchenne's muscular dystrophy, and cystic fibrosis. Similar work has begun for colonic polyposis and its associated cancer of the bowel, Alzheimer's disease, manic depressive affective disease, and many others.

Not only has this approach proved valuable in diagnosis, but the sequences revealed by RFLPs provide a starting point for the isolation of the relevant disease gene itself. Unfortunately, because the present methods of moving along a genome to get to a disease-related gene are laborious and tedious, they cannot be used easily to span the frequently enormous distances from the marker to the disease locus. For instance, a chromosomal walk of several hundred contiguous kilobases (1 kilobase = 1,000 bases) has been achieved at great effort in only a small number of laboratories. Since an average human chromosome may contain several hundred thousand kilobases, there is an urgent need for more efficient methods to move more rapidly from one gene locus to another on a chromosome.

Once a candidate gene is isolated, a number of methods are used to establish whether it is indeed the targeted disease-related gene. Success in isolating the specific disease-related gene then permits simpler and more precise detection of abnormal forms of the gene, the identification of the gene product, and an educated guess about its function through structural and evolutionary studies. The candidate gene must be demonstrated to have an altered structure and function in patients with the disease under study.

The most impressive of all the achievements of reverse genetics has been the recent identification and isolation of the gene responsible for cystic fibrosis. The potential problems sure to be encountered frequently in similar reverse genetic projects are illustrated by the problems associated with efforts to isolate the Huntington's disease gene. It has been more than seven years since the first published report of linkage of the RFLP marker to Huntington's disease, but even with the collaborative efforts of several large groups, consuming hundreds of person-years of work, the disease gene has remained elusive. The availability of a more thoroughly saturated linkage map, together with other technical advances, will speed progress.

Because a linkage map of the human genome with a resolution of 3–5 centimorgans is nearing completion, genetic diseases will soon be mapped much more quickly. Such a detailed map will define the relative positions of all known genetic loci, identify the RFLPs linked to most and perhaps all human genetic diseases, and provide a starting position for genomic walks to the relevant disease loci themselves. On the other hand, a map will provide limited information on the vast regions of the genome thought to have little to do with genes or their expression. These regions include the so-called junk DNA or selfish DNA found in enormous amounts in the genome, sequences whose only purpose, some believe, may be to replicate themselves. Whether these regions are worthy of extensive characterization is at issue. Although it would be inappropriate at this stage of our ignorance to exclude them on the assumption that they serve no useful genetic function, regions containing authentic genes are almost certainly more appropriate for an initial detailed study.

A physical map is similar, but not identical, to a genetic linkage map, because genetic distances defined by recombinational events do not always reflect precisely the physical distance between markers. A linkage map therefore represents a view of a genome through a distorting lens, and the task of designing a walk from closely linked markers is made difficult by not knowing how far to walk and how to recognize the target when it has been reached.

The Human Genome Project

One can now envision the structural and sequence characterization of the entire human genome. A number of large-scale studies have already begun. Three different stages or approaches can be envisioned:

development of a physical map of each chromosome and the isolation of overlapping fragments that span it; sequence determination of particularly important or interesting regions; and systematic sequence determination without prior evidence on function.

The first approach, a fine-structure physical map and the isolation of cloned, overlapping restriction fragments, is an extension of currently available techniques already used successfully to study the genomes of the bacterium *Escherichia coli* and the nematode worm *Caenorhabditis elegans*. The entire *E. coli* genome, of 4×10^6 nucleotide pairs, has been ordered and aligned as a contiguous set of clones. The size of the nematode genome, approximately 80×10^6 nucleotide pairs, is approximately the same as that of a small human chromosome or a fragment of a larger chromosome, implying that such an approach may already be technically feasible for large regions of the human genome. The most important feature of this approach is that it would provide not only map information but also a new kind of reagent—a collection of cloned fragments that overlap each other in a known order and that together span an entire chromosome or large portion of a chromosome. Since the relationship of these clones to all known disease loci on that chromosome would be known, it would not be nearly as difficult to walk from a linked RFLP to a new disease locus.

As sets of contiguous fragments become available for large regions of the genome, disease loci will quickly become available as pure clones, since the most difficult and rate-limiting step—long genomic walks—will have been eliminated. One would walk to the freezer to get a disease gene rather than try to walk down the chromosome. If ordered libraries had been available when the RFLP-linked probes were first identified for cystic fibrosis, Duchenne's muscular dystrophy, fragile-X syndrome, Huntington's disease, Alzheimer's disease, or manic depressive disease, all of these genes would probably already be in hand. With such genetic libraries available, the time required to go from the identification of an "anonymous" genetic marker to the successful isolation of the cystic fibrosis gene itself would have been weeks to months rather than the years that it did require. The same will be true of the many other RFLP-linked disease loci to be identified in the coming years.

The isolation of specific target genes or regulatory regions will be followed by nucleotide sequence determination. Since such regions are likely to be reasonably small, there seems to be no barrier to rapid sequence characterization of many important genes with resulting improved understanding of their role in disease. However, some loci, such as that for Duchenne's muscular dystrophy, are so long, in the range of several million nucleotides, that they present more formidable sequencing undertakings.

The second major path to characterizing the human genome is the nucleotide sequence determination of selected regions known to contain genes and disease loci of special interest or importance. Since most of the human genome does not encode any recognized useful functions, many argue that it would be more efficient to concentrate our efforts first on those regions of the human genome, no more than a small percentage of the total, known to contain genes.

The third approach would be to make an immediate start on a total sequence determination of the entire genome, without any prior or weighted selection of target regions. However, most agree that present sequencing methods are not efficient enough to warrant this approach. The task is akin to determining the arrangement of every letter of every word in ten complete sets of edition 25 of the *Encyclopaedia Britannica* and, using today's methods, would probably cost hundreds of billions of dollars. The combination of an order-of-magnitude improvement in sequencing technology and data management, together with a well-organized international collaborative effort, will begin to make direct sequencing more and more attractive.

Technical problems aside, a sequencing project per se will eventually isolate, map, and characterize genes and intergene sequences that will be useful as direct or linked probes for disease detection and possibly disease treatment. Direct, untargeted sequencing will lead to clinically useful reagents only after it is well under way. Interpretations of sequence information will also require a great deal of sequence comparison, not only between individual human beings, to distinguish between polymorphisms and disease-related mutations, but also between species, to identify the significance of sequences conserved in evolution.

The Human Genome Project is now well under way. With the current explosion of technical and conceptual innovations, one of its early goals, completing a fine-structure physical map within five years, is within reach. Both the U.S. National Institutes of Health and the U.S. Department of Energy have committed sizable and rapidly increasing funds to different aspects of the project, and they anticipate that approximately $3 billion will have been devoted to it over the coming fifteen years in the United States alone. Substantial projects are under way in Europe, Japan, and the Soviet Union, and efforts are being made by all of these individual programs to coordinate their efforts, under the aegis of the privately funded international Human Genome Organization.

The flow of funds is encouraging to the project's proponents, but it is a cause of concern to other scientists who expect that other fields of biomedical research will suffer corresponding reductions in funding in this era of constrained national resources and a waning federal commitment to the support of some parts of the American biomedical research apparatus. There is indeed some justification for these fears.

Implications for Clinical Medicine

Whether the information comes from the genome project or from other studies, a better knowledge of the structure of the human genome will affect medicine in several ways. These include a more complete understanding of the pathogenesis of disease, simplified diagnostic and screening procedures, earlier diagnosis, and sometimes more effective treatment (when mutant genes are detected before cell or tissue damage takes place). Some medical investigators exaggerate the early prospects for translating complete nucleotide sequence information into efficient treatment. For example, there are few genes understood as well as those of the hemoglobin proteins; and yet progress toward truly effective therapy for diseases affecting hemoglobin production has not moved at nearly the same pace. The isolation and characterization of genes will continue to have a far greater impact on detection, diagnosis, screening, and genetic counseling than on therapy.

As with most other medical and scientific advances, the benefits derived from an extensive characterization of the human genome will come with costs and problems—financial, scientific, and societal. A great deal of important basic scientific information will certainly be collected; but the magnitude and the timing of the medical benefits are less clear. So much new knowledge of genetic diseases, and so many new screening and detection programs, will magnify the problems already created by increasing demands on our limited resources and imagination.

One problem involves the choice of diseases and target populations for large-scale detection programs. As effective detection methods become available for major, common, high-burden diseases, such as cancer, cardiovascular diseases, and degenerative disorders including Alzheimer's and Parkinson's diseases, effective and desirable screening and diagnostic programs will be instituted, especially for disorders in which early treatment and genetic counseling would be beneficial. Our society already supports virtually universal and mandatory newborn screening programs for a number of such disorders, including phenylketonuria (PKU) and galactosemia, both causes of severe mental retardation if not detected shortly after birth. Voluntary preventive programs aimed at specific target diseases also exist, such as that for Tay-Sachs disease, among Ashkenazic Jewish populations, which leads to death early in childhood.

However, diseases for which there is no effective therapy pose many difficulties for patients and health systems alike, as illustrated by the untreatable Huntington's disease, a terrible dominant genetic degenerative neurological disease that strikes unsuspecting adults usually long after they have passed the disease gene to their children. RFLP-linked diagnostic procedures are already feasible for some families. However, no medical purpose is served by knowing whether or not one is carrying this gene, and hence is fated to eventual slow neurological deterioration and death. Many who do not yet know whether they are affected choose not to know, and to live in blissful ignorance, hoping for the best or contenting themselves with the knowledge that many years of useful life are probably granted even to those afflicted with the lethal gene. On the other hand, the test can be used for reproductive counseling, to prevent the birth of children with the defect.

Many other severe diseases will become susceptible to detection and control but will become "orphaned" or ignored because they are simply too rare to seem to justify large-scale detection programs. Genetic screening will then be limited to families in which the disease has already made its appearance.

For some common, high-burden diseases, such as coronary artery disease, hypercholesterolemia, cystic fibrosis, manic depressive (bipolar) disease, and Alzheimer's disease, large-scale population screening would be helpful, but it probably cannot be carried out until gene-specific probes become available. The present RFLP analysis is simply too time-consuming and labor-intensive to be used to examine very large populations. Even so, the costs and logistic problems associated with such widespread programs will be enormous.

It will be vital to determine which diseases should be included in screening programs, and in what order, and whether there are any indications for making such programs mandatory. With increasing pressure for the development of a more equitable health care delivery system, and with a more pervasive role for government and the insurance industry in financing universal health care, the pressure for population-wide screening programs for growing numbers of disorders and defects is inevitable. Since part of the effort will be devoted to prenatal diagnosis, and since the most feasible medical response to the presence of a prenatal defect is the termination of pregnancy, the flood of new prenatal genetic information is likely to exacerbate the already low quality of the discourse on this issue.

At the same time, for some diseases effective treatments will provide powerful positive influences to balance the potential economic and ethical hazards of identification of patients and carriers of gene defects. Mandatory screening for PKU and galactosemia has been accepted because the threat to personal autonomy seems negligible when balanced against the good that can be done by early detection of these treatable diseases.

Instituting broad, genetically based screening programs is not necessarily an obvious good, and one does not have to point to the most egregious abuses of genetic or "pseudo-genetic" information to recognize its potential for misadventure or outright evil.[4] In Nazi Germany some "scientists"—psychiatrists, geneticists, and anthropologists—as much as the politicians, pioneered and fostered the concepts and the principles that rationalized the racial underpinnings of the Nazi extermination programs.[5] There have been less sinister instances in which apparently reliable scientific information nevertheless led to inappropriate social policy. For example, many sickle cell carriers with no increase in health risks experienced problems with employment and insurability following the establishment of screening programs in the 1970s in the United States. Disorders such as Huntington's disease, schizophrenia, and bipolar disease illustrate the point that genetic screening programs might not be universally desired or accepted when there is no effective therapy for the disease.

As genetic mapping leads to the development of many new genetic diagnostic procedures, the financial costs will also present problems. Insurance companies and other third-party carriers often do not now reimburse patients fully for heterozygote detection procedures or for genetic counseling. With the likelihood that genetic diagnosis will become feasible for the most common and severe diseases of our society, both public and private nonprofit and for-profit institutions can anticipate a massive load of new diagnostic tests and screening programs.

As with other medical information, the ability to keep genetic diagnostic information confidential will be threatened. Again, there is little conceptually new here, except for the magnitude of the problem and its potential for abuse. Issues of mandatory genetic screening prior to employment and to health or life insurance coverage represent some of the potentially most difficult social problems for the future predictive power of molecular human genetics.

Finally, there will be an exacerbation of a variety of medico-legal issues. In a field as rapidly developing as human genetics, services will not be distributed uniformly, and inadequately trained or unprepared health care providers will be faced with a flood of new techniques and procedures. It is inevitable that questions will be raised of breached confidentiality, genetic damage, and wrongful life, as well as medical negligence. Difficult issues of jurisprudence will surface with far greater frequency than at present. Though not unique to new molecular genetic techniques, these issues will grow in importance with the explosive increase in the development and use of these tools.

Somatic Cell Gene Therapy

Even with the application of modern tools of molecular biology, there continue to be major temporal and conceptual gaps between an understanding of the mechanisms of disease and its effective treatment, but they will close with increasing speed. In a powerful new approach, called gene therapy,[6] a nondefective gene is introduced into cells from patients to correct a genetic defect. While gene therapy still poses many technical, medical, and ethical problems, most of the uncertainties about its feasibility and desirability have dissipated, and its experimental use for otherwise intractable disorders has begun.

In the most common current approach to gene therapy, the missing genetic information is introduced into suitable defective target cells cultured from the patient, and modified cells are then introduced into the patient. This restoration of genetic functions to the body cells (somatic cells) of an individual will not affect future generations. Modified, defective retroviruses (lacking a sequence required for pathogenicity) have been used widely as vectors for the foreign sequences, since retroviruses have evolved to do precisely the job demanded of a therapeutic agent: to introduce genes efficiently and stably into cells without damaging them.

By means of these agents genetic defects have been successfully corrected in human and other mammalian cells in culture. A number of human disease-related genes have also been made to function adequately in vivo, particularly in the bone marrow of mice and, more recently, of primates. However, the expression of such foreign genes, called transgenes, from retrovirus vectors has often been inefficient and transient. Progress is being made in solving some of these technical problems.

There are also promising approaches to the genetic modification of organs other than the bone marrow. Possibly the simplest target organ would be the skin, since its cells are easily obtained from an individual, cultured, manipulated in vitro, and returned to the patient. These cells might provide circulating functions, such as insulin and antihemophilic blood coagulation proteins, even though their location is abnormal for this purpose. The endothelial cells lining the blood vessels may also be useful for this approach. There is promise as well that genetically modified liver cells may restore missing metabolic function to the liver, thus eliminating, for example, causes of elevated blood cholesterol levels, or the defect in PKU.

Therapy of Cancer and Central Nervous System Disorders

A great deal has been learned about genes involved in critical events regulating cell growth and differentiation. When some of these normal cellular genes, called *proto-oncogenes*, become mutated or are expressed inappropriately, they are converted into active tumor-producing genes, called *oncogenes*. Some oncogenes act in a dominant way, transforming normal cells to cancer cells. These are called *transforming oncogenes*. Other, recessive, genes, whose normal role in the cell is to exert a growth-suppressing function, act as oncogenes only when both copies become inactive. These are termed *cancer suppressor genes*. Recent work has shown that it may be possible to treat the tumors resulting from absence of a cancer suppressor gene, such as the common eye tumor of children called retinoblastoma, by introducing a normal copy of the gene into the tumor cells, which contain only abnormal copies.[7] Vector-mediated delivery of suppressor genes may eventually play a significant role in cancer therapy.

It is more difficult to envision a genetic approach to the treatment of tumors caused by the dominantly acting transforming oncogenes, since it would seem to be necessary to turn them off or otherwise inactivate them. However, techniques for selectively disrupting a specific gene are beginning to be feasible. Though they are still inefficient, the door seems to be opening slowly to targeted inactivation of dominantly acting oncogenes and even of other dominant genes, like that responsible for Huntington's disease.

Similar genetic approaches are being taken to some disorders of the central nervous system, although serious conceptual and technical problems must be solved. The presumed target cells, the neurons, are relatively inaccessible, and since they do not multiply they are not susceptible to infection with retrovirus vectors, which require active cell replication for infection. One potential solution is the use of non-neuronal cells infected in culture and subsequently implanted into the central nervous system, where they release useful products. This indirect method has permitted partial correction of the cell abnormalities and aberrant behavior of rats with defects similar to those found in human patients with Alzheimer's or Parkinson's disease.

For the treatment of many other, more global disorders of the brain, it is vital to develop vectors that can transfer foreign genetic material into nonmultiplying cells.[8] Herpes simplex virus type 1, the ubiquitous and generally benign human virus that causes "cold sores," is an attractive candidate, for a number of reasons. It is capable of establishing a latent, lifelong, noninjurious infection in neurons; it does not seem to carry an oncogene; and it does not integrate into the host genome, thereby reducing the potential for genetic damage in the cells. Furthermore, many of its own genes are dispensable for virus growth, thereby making room for replacement by foreign genes. Such genes incorporated into this virus have already been expressed in cultured cells.

Muscular dystrophy in humans has been found to result from mutations in the gene for the muscle protein called dystrophin. The gene has been cloned and characterized, and very useful models have been found in the mouse and the dog. Furthermore, implantation of normal muscle cells into muscle of the affected mouse leads to improvement, suggesting that genetic modification of even a portion of the defective cells in vivo may result in a beneficial effect on muscle function. With mechanisms of muscle-specific gene expression becoming better understood, this previously untreatable disease may become an important target for gene therapy.

Finally, the successful isolation of the gene responsible for cystic fibrosis makes this disorder an attractive target for gene therapy. Cystic fibrosis is the most common lethal genetic disease in the United States, occurring in approximately one of every thousand live births. It is associated with a reduced chloride secretion through membranes of cells in the sweat gland ducts, the intestine, and the airway, leading to more viscous mucus. The disease cannot be adequately treated at present, and the problems of delivering the normal gene to the relevant cells are formidable. However, they are probably not insurmountable.

Germ Line Manipulation

Some diseases might eventually be treated permanently and heritably by modification of the genes in egg and sperm cells in a process called germ line therapy. The process has already been carried out, albeit with some technical difficulty, in lower mammals. Many of the requisite techniques are well developed, and in mice it is routinely possible to introduce foreign genes into fertilized mouse eggs and to reimplant them into pseudo-pregnant foster mothers. The resultant genetically modified "transgenic" mice carry and express such genes and pass them on to their own offspring. If this approach were to become technically and ethically feasible in human beings, a single genetic modification might prevent disease not only in specific individuals but also in all of their progeny.

Germ line genetic modifications of human beings has been criticized more than somatic cell therapy, on both technical and moral grounds. The efficiency of production of transgenic mice is low, with many more failures than successes and an occasional new defect—obviously an impermissible situation in work with human beings. In addition, many critics consider it an unnecessary technology, since selective abortion could accomplish the same goal for many diseases with existing techniques—although the use of abortion is far from a universally accepted response to the detection of a genetic disorder. Alternatively, it will soon be possible to diagnose many diseases in single cells isolated from very early human embryos after in vitro fertilization. Only unaffected embryos might then be returned to the mother, thereby ensuring that the child will not be affected with the disease in question. Of course, this approach has severe limitations, since in vitro fertilization is extremely expensive and therefore will be inequitably distributed.

Many other critics consider that the designed alteration of human beings and their progeny represents a capacity that ought to be forbidden on ethical grounds. Once we start, we may be tempted to extend this kind of manipulation beyond the prevention of disease into the enhancement or modification of traits whose value may be much less clear. Critics argue that we are not wise enough to know which, if any, human traits can be modified without dire social or evolutionary consequences.

Despite medical justification for germ line change to modify or eliminate a gene that all can agree is deleterious and undesirable, the concept has generally been dismissed on technical as well as ethical grounds. But the advent of more effective scientific tools will add urgency to this discussion. It is very likely that the issues will come to be so thoroughly discussed and examined that consensus will be achieved, just as has occurred with somatic cell manipulation.

Social Costs and Goals

The acquisition and use of genetic information will have not only economic but also social costs. Issues dealing with our genetic nature seem to be especially threatening to us, possibly because they strike so closely at our individual uniqueness and at how we view our origins, future, and relationship with the rest of the cosmos. Fortunately, there are limitations to the extent to which we can redesign ourselves genetically, especially our complex mental traits. But there also are, or ought to be, ethical and public policy boundaries to our pursuit of genetic knowledge and the correction of genetic disease. The explosion of knowledge now occurring in human genetics will multiply the opportunities for both beneficence and mischief.

The specific goal of the Human Genome Project—the total structural characterization of the human and related genomes—is a difficult but straightforward technological undertaking. It will be done, and before very long. But although many of the mysteries of gene organization and expression will be difficult to unravel, the role of genes in human development and illness will be increasingly understood.

We shall come to understand which genes are responsible for defects in development and metabolism, tumor development, aging, neuropsychiatric disorders, and many other illnesses. As a result, we shall be capable of undertaking screening and detection programs for many disorders, which often lack effective therapy. We shall also be able to use genetic tests to identify individuals for medico-legal and forensic purposes. These applications raise troublesome questions, however, regarding the involuntary application of predictive genetic techniques in other areas, such as employment and medical or life insurance settings, where the rights of individuals, employers, and insurance companies will inevitably clash.

The prospects and promise for therapy are less complicated and more defensible. As moral agents most of us can agree that certain human conditions should be eliminated when possible because they cause suffering and limit our freedoms. We accept as a blessing their amelioration through drug treatments or surgery. As long as there is general agreement on what constitutes "disease," there are few dilemmas. However, the distinction from deviations within the normal range will almost certainly become more difficult as genetic determinants for many multigenic traits come to be recognized.

A remarkable change has occurred in our attitudes toward gene therapy. Somatic cell therapy, so revolutionary a concept only a few years ago, is now taken for granted. The conceptual revolution has already taken place. Its goals are consistent with the noblest traditions of medicine and apparently pose no new ethical problems. All major religious, governmental, and public policy bodies that have examined gene therapy, including the Roman Catholic Church, the World Council of Churches, the American Council of Churches, the U.S. Presidential Commission for the Study of Ethical Problems in Medicine and Biomedical Research, and the Parliamentary Assembly of the Council of Europe, have concluded that the concept is ethically acceptable.

The goal of modern human genetics is to understand not only our disorders but also the genetic bases for those qualities that distinguish us from other species: our capacity to think and wonder, to foresee and imagine, to communicate complex messages, to create and pass on learning and memories. Surprisingly, the genomes of the human and of higher nonhuman primates have been found to differ in sequence by perhaps as little as 1 percent. Within this small set of differences it will be possible eventually to understand many aspects of the genetic mechanisms that determine uniquely human traits, as staggering as that challenge may be.

Nevertheless, while the relatively limited differences between human and chimpanzee genomes will eventually be determined, the sequence of the genes will not tell us all there is to know about our peculiarly human or individual characteristics; the journey from a description of gene structure to an understanding of the development of the complex wiring patterns of the human brain, its cellular communication networks, and its physiological events will be a long and arduous one.

Conclusion

The product of inquiry ought to be value-free, but one wonders if some kinds of scientific knowledge lead so inexorably to misadventure and abuse that they ought not to be pursued or, once achieved, ought not to be applied. The increasing acquisition and application of genetic knowledge could pose this ethical quandary. The relief of human suffering through genetics is certainly a good and should be pursued. But the possibility of frivolous tampering with human qualities, the arbitrary enhancement of some traits outside the range of diseases requiring correction, raises fears about an expanding application of modern molecular genetics to the human condition.

These fears are intensified by past errors and abuses in the interaction of genetics with social policy. Examples include the American misuse of pseudo-genetic information to support restrictive immigration policies during the 1920s and 1930s; the more recent episodes of unjust and scientifically unsound insurance and hiring policies for heterozygous carriers of the sickle cell defect; the role of German psychiatrists, anthropologists, and geneticists in the establishment of a "scientific" basis for the Nazi racial policies; the destruction of Soviet genetics by ideology-inspired scientific nonsense during the Lysenko era; the "evidence" provided by "eminent" physical and social scientists of the genetic inferiority of blacks; and, conversely, objections to any study of the genetic contributions to human traits such as intelligence. These episodes illustrate the uncomfortable susceptibility of genetics to political abuse and the tendency of evil and self-serving political ideologies and social forces to exploit distorted science.

Even in the absence of malevolence, genetic information can be distorted for political purposes. The abysmal quality of the debate over abortion in the United States revolves around pseudo-scientific arguments fabricated to buttress fundamentally theological and philosophical stances. Not to be outdone at sophistry, others misuse the authority of science to deny the existence of difficult ethical dilemmas worthy of serious debate. But as in the past, problems are more likely to arise in the future from our ignorance and our inadequate social institutions than from our knowledge. Enlightened knowledge of human characteristics and of our relationship with all around us cannot in itself be dangerous and should not be forbidden.

But some kinds of pursuit or application of scientific knowledge, whether genetic or otherwise, are likely to be so far beyond ethical bounds as to be impermissible. We have seen this in the Nazi medical experiments, in the Tuskeegee studies in this country in which effective treatment with penicillin was withheld from a group of black patients with syphilis to determine the natural history of disease, and in the denial of insurance coverage to carriers of the sickle cell trait. There have been many other misguided applications, and there will be more.

The powerful new genetic tools and knowledge that will be generated by the human genome initiative, and by deliberate genetic intervention in human beings, must not be repressed. Both approaches promise to lead to immense good. But that good will not come without costs—scientific, social, and ethical. We have an opportunity to develop and use these new powers and tools carefully and well, or bluntly and clumsily. We shall probably do both.

Notes

1. Office of Technology Assessment, *Mapping Our Genes: Genome Projects—How Big, How Fast?* (Washington, D.C.: OTA, 1988; Baltimore: Johns Hopkins University Press, 1988).

2. V. A. McKusick, *Mendelian Inheritance in Man* (Baltimore: Johns Hopkins University Press, 1988).

3. J. B. Stanbury, J. B. Wyngaarden, D. S. Fredrickson, J. L. Goldstein, and M. S. Brown, *The Metabolic Basis of Inherited Disease* (New York: McGraw-Hill, 1988).

4. D. J. Kevles, *In the Name of Eugenics* (New York: Alfred A. Knopf, 1985).

5. B. Müller-Hill, *Murderous Science* (Oxford: Oxford University Press, 1988).

6. T. Friedmann, "Progress toward Human Gene Therapy," *Science* 244 (1989): 1275.

7. H.-J. S. Huang et al., "Suppression of the Neoplastic Phenotype by Re-

placement of the Human Retinoblastoma Gene Product in Retinoblastoma and Osteosarcoma Cells," *Science* 242 (1988): 1563.

8.   M. B. Rosenberg et al., "Grafting of Genetically Modified Cells to the Damaged Brain: Restorative Effects of NGF Gene Replacement," *Science* 242 (1988): 1575.

10

Neuroscience

James H. Schwartz

Neuroscience has always been interdisciplinary, combining diverse fields of research—anatomy, chemistry, physiology, and clinical studies. Recently it has added molecular biology. Its techniques and the approaches of modern cell biology are now greatly increasing our scientific understanding of the brain and its disorders.

Although scientific knowledge of the brain, and perhaps of the mind as well, has been greatly augmented, I have reservations about orthodox Panglossian ideas on the philosophical and social consequences of this understanding. These reservations can be introduced by two comments that, while not contradictory, represent two very different social perspectives on this progress in our scientific understanding. One comment is by Jacques Loeb, a physiologist whose optimistic outlook profoundly influenced twentieth-century biology: "It was perhaps not the least of Darwin's services to science that the boldness of his conceptions gave to the experimental biologist courage to enter upon the attempt of controlling at will the life-phenomena of animals, and of bringing about effects which cannot be expected in Nature."[1] Loeb's emphasis is that progress in science enables the biologist to control nature. Robert M. Young, a social historian of science, presents a more cautious view: "One could indeed tell a chilling tale about the theory of evolution in its various forms as the greatest single blow to man's self-esteem, a tale which rests on the fact that evolution includes mind in the course of natural law."[2]

In this chapter I shall first review the new, scientifically exciting and medically promising knowledge of the neurosciences in its historical context. Then, at the end of the chapter, I shall attempt to evaluate the impact of this knowledge, both scientific and social.

A Brief History

About two hundred years ago almost everybody finally became convinced that the mind is in the head and not in the heart. It is with the German neuroanatomist Franz Joseph Gall that the modern period of neuroscience begins:

> The body of man consists of matter: hence it is subject to all the laws of matter . . . The comparison of man with other beings (not merely animals, but also plants and minerals) must be admitted, and cannot be repugnant to our feelings, because he participates [in] their properties . . . Man unites all physical, chemical, vegetable, and animal laws; and it is only by particular faculties that he acquires the character of humanity. The brain is exclusively the organ of the feeling and intellectual faculties . . . Thus all concurs to prove that the brain must be considered as the organ of the moral sentiments and intellectual faculties.[3]

Gall's work was censured in 1802 by the Austrian government, and three years later he was expelled from Vienna because of the subversive and materialistic implications of his theories. For the same reasons he was welcome in post-Revolutionary Paris, where he died in 1828, a savant so celebrated that a memorial medal was issued by the French mint to commemorate his death.

Until recently, Gall's reputation was poor because his writings, and the proselytizing of his disciples, led to the enormous popularity of phrenology. Phrenology was based on two theoretical principles. First, the brain is made up of from twenty-five to nearly one hundred discrete organs that each control a particular faculty: either simple—for example, combativeness, alimentiveness (appetite for sustenance), benevolence, and sight—or complex—for example, philoprogenitiveness (parental tenderness), sublimity (conception of grandeur), marvellousness (wonder and credulity), and ideality (imagination).[4] Second, the size of these organs and their contribution to temperament differ in different people. The practice of phrenology depended upon two consequences of these principles. Temperament and its disorders could be diagnosed by examining the cranium, since a prominent organ enlarges the skull over the particular region of the brain in which it is situated. Treatment followed from the belief that, as with muscle, exercise could enlarge a desirable organ of the brain.

Though wrong in detail, Gall's view of the brain was right in principle. His work persuaded many scientists, most physicians, and the educated public that the brain is the organ of the mind, and that mental functions are localized to specific parts of the brain. This view contradicted the notion that the brain produces a glandular or spiritual soup—an idea that stemmed from Galen, the second-century Greco-Roman physician whose doctrines had dominated medicine from antiquity to the eighteenth century.

For the most part, practicing physicians came from the emerging middle class. They found phrenology attractive because of Gall's liberating principle: that the exercise of specific brain centers could improve the faculties that they controlled. Thus, by insisting on the material basis of mind, as well as on its capacity for self-improvement, Gall explained "scientifically" how members of the rising middle class could break the class limitations imposed by church and aristocracy: they could rise and improve their mental capacities by using them. Virtue and quality were not necessarily co-linear with noble birth, but rather could be bettered through hard work.[5] Furthermore, if mental functions were localized, the processes underlying them could be studied, understood, and ultimately controlled.

Localization of brain function derived support from clinical observations, starting with the great nineteenth-century neurologists Pierre Paul Broca and Carl Wernicke. Both men made their discoveries on brains of patients with aphasias (disorders of language) who had suffered strokes and later died. In 1861, Broca found that damage to a region of the left frontal cortex, a part of the brain now called Broca's area, resulted in the inability to speak without loss of intellectual function or the capacity to understand and read language. In 1876, Wernicke observed patients who were able to speak but could not comprehend language. The brains of these patients were damaged in a part of the left temporal cortex, now called Wernicke's area. While unscientific cranioscopic phrenology continued to be practiced as late as the beginning of World War I,[6] other neurologists produced specific and useful maps of function like the somatotopic map by the Canadian neurosurgeon Wilder Penfield showing the motor cortex as a homunculus (Figure 10.1).[7]

Despite this persuasive clinical evidence, localization of function to specific regions of the brain was opposed by some physiologists and psychologists,[8] starting with Marie-Jean Pierre Flourens in the second quarter of the nineteenth century. To discredit the evidence linking a specific area with a particular function, they used experiments in animals in which specific areas of the brain were surgically removed. Often the function attributed to the injured part was not entirely lost; conversely, damage to other areas of the brain might result in loss of a function attributed to an uninjured part. These experiments were unreliable because accurate placement of lesions was difficult to achieve, and the conditions of surgery were primitive and unsterile. In addition, the intact parts of a damaged brain can compensate for loss of function to some degree—though this usually is a small effect, in part because the mature brain has only a limited capacity to regenerate.

Even though this evidence against localization of brain function was beset with serious technical problems, the opposing theory (called

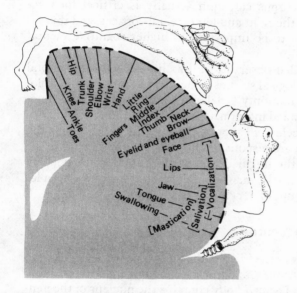

Figure 10.1 Representation of the areas of the right cerebral cortex devoted to various muscle groups on the opposite side of the body. The so-called homunculus reveals that larger cortical areas are involved with the tongue, face, and fingers than with the rest of the body. Clearly, in addition to an ordered sequence of representation of body parts, there is a disproportionate amount of cortex devoted to muscles concerned with skilled movement. The map was obtained by direct electrical stimulus of the motor cortex (precentral gyrus) in awake patients by Wilder Penfield during operations for the surgical treatment of epilepsy. (After W. Penfield and T. Rasmussen, *The Cerebral Cortex of Man: A Clinical Study of Localization and Function* [New York: Macmillan, 1950]; from E. R. Kandel, J. H. Schwartz, and T. Jessell, *Principles of Neural Science*, 3d ed. [New York: Elsevier, 1991].)

the aggregate field view) dominated psychology in the first half of the twentieth century, and was reflected in America by the mechanistic views of Jacques Loeb and the behaviorism of James B. Watson and Burrhus F. Skinner.[9] In part the aggregate field view was an implication of the "black box" approach that these psychologists and their disciples took to the nervous system. Concerned with the external stimulus and the animal's response, they considered the structure and function of the intervening organ of the mind, if not unimportant, at least unknowable. Karl S. Lashley postulated that the mass of brain tissue, not specific centers in the brain, is critical to function. As late as 1950 he wrote that the function of the corpus callosum (which connects the cerebral hemispheres) "must be mainly mechanical . . . i.e., to keep the hemispheres from sagging."[10] (The corpus callosum actually is critical for integrating the functions of the right and left cerebral cortices, and this integration is now known to be important to higher cognitive mental processes.)[11]

Nevertheless, detailed observations by clinical neurologists and neurosurgeons, morphological studies by neuroanatomists and histologists (notably Santiago Ramón y Cajal and Camillo Golgi), and functional studies by sensory and motor neurophysiologists (notably Charles Sherrington and John Eccles), all showed that specific nerve cells in the central nervous system are responsible for carrying out particular aspects of behavior. Like the linguistic and motor functions mapped by Broca, Wernicke, and Penfield, many neural faculties have now been localized to identified groups of nerve cells in the brain.[12]

An Overview of the Neuron

The functional unit of the brain is the neuron or nerve cell (Figure 10.2). Neurons typically have four parts: a cell body, dendrites, an axon, and synaptic terminals. The *cell body* contains the nucleus of the neuron and therefore the genes that encode the information for synthesizing the neuron's proteins. Proteins are synthesized in the cytoplasm surrounding the nucleus and not in the other parts of the nerve cell. Cytoplasm is a gel composed of water-soluble proteins, salts, and other low-molecular-weight molecules, and some membrane-bound subcellular organelles. In addition, cytoplasm contains several kinds of protein fibers that make up the cytoskeleton, which is responsible for cell shape. Typically many *dendrites*, which are thin extensions of the cell body that act rather like antennae, receive inputs from other nerve cells, or from physical stimulation of sensory organs such as pain receptors in the skin. Even though dendrites are specialized for receiving inputs, reception can take place at points of contact in all parts of the neuron.

Wherever received, inputs are transformed into one of two types of signals: ionic or biochemical.

I shall describe ionic signaling first; signal transduction mechanisms (biochemical signaling) will be discussed later. An *axon*, a thin, tubular process of the cell body usually much longer than dendrites, conducts the nerve impulse. Stiff cytoskeletal fibers in the cytoplasm maintain the thin, tubular shape of both dendrites and axon. Because cytoplasm is composed chiefly of water and charged molecules, it readily allows movement of ions and is therefore a good conductor of electrical current. At its end, the axon branches into many fine processes called *neurites* or *terminals,* which end at specialized points of contact with target cells, called *synapses* (Figure 10.3). Nerve cells communicate with one another by means of synaptic transmission. The presynaptic neuron releases one or more neurotransmitters from its point of contact. Neurotransmitters then act on specialized proteins called *receptors* in the receiving or postsynaptic cells.[13]

Neurotransmitters are either small molecules or somewhat larger peptides. Their release occurs when a wave of decreased electrical potential, called *depolarization,* sweeps down the membrane of the axon and reaches the terminals. Depolarization is usually initiated at a specialized region of the axon close to the cell body, as a result of multiple excitatory and inhibitory synaptic connections of other nerve cells on the dendrites. Excitatory (depolarizing) and inhibitory (hyperpolarizing) inputs have opposite effects on the electrical potential of the membrane. If the net depolarization is sufficient, a nerve impulse, the *action potential* (a self-propagating depolarization), results and is propagated along the axon toward the nerve's terminals.

When the depolarization reaches the terminal it opens voltage-sensitive ion channels, large protein molecules which form a pore through the membrane. When opened, the pore allows the influx of calcium ions. These ions cause many synaptic vesicles—tiny membrane-bound spheres filled with neurotransmitter that cluster at the synaptic membrane in the terminals—each to release its quantum of transmitter from the cell.[14] Once released, the transmitter molecules diffuse across the narrow space between the presynaptic terminal and the postsynaptic cell to interact with postsynaptic receptors on other neurons, or on target cells such as muscles or glands. In ionic signaling, this interaction results in the influx of ions into the postsynaptic target, thereby changing its electrical properties. As an example, the resulting depolarization of muscle cells causes them to contract.

The process of synaptic transmission is regulated in several distinctive ways, both presynaptic and postsynaptic. Presynaptic mechanisms may change the amounts of neurotransmitter produced, alter the availability of the vesicles that contain it for release, or modulate the effects of depolarization, usually through regulating the ion channels that control entry of calcium ion. Postsynaptic mechanisms alter synaptic transmission by changing the functional properties of receptors.

The study of the molecular basis of synaptic transmission was made possible by technologies that were essentially spin-offs from military research and development for World War II. Development of sophisticated electronics was needed for work with glass microelectrodes, which are central to electrophysiological and biophysical studies of the nerve cell; for the electron microscope, first used to examine the anatomical fine structure of nerve cells in the early 1950s; and for perfecting methods and equipment for separating the biochemical constituents of neurons.[15] Similarly, modern molecular studies of neurons as well as all other cells are inconceivable without radioactively labeled chemicals, which began to be generally available for biological and medical research by 1950. They have been invaluable reagents for studying metabolic pathways and for permitting the detection of trace components.

Ion channels, synaptic vesicles, peptide neurotransmitters, and re-

Figure 10.2   Main features of a typical neuron. *A.* Drawing of a pyramidal neuron by the Spanish neuroanatomist Santiago Ramón y Cajal. These cells are the principal type of neuron in the cerebral cortex. *B.* Schematic nerve cell showing the main features of a typical neuron: cell body, dendrites, axon, and terminals. Points of synaptic input from other nerve cells are indicated by triangles on dendrites and cell body: excitatory presynaptic terminals are white, and inhibitory terminals are black. Axon terminals of one neuron make synaptic contact with the dendrites or cell body of another neuron. Thus, dendrites might receive incoming signals from hundreds or even thousands of other neurons. The axon can be seen to emerge from the cell body. Axons vary greatly in length, some extending for more than 1 meter. Most axons in the central nervous system are very thin compared with the diameter of the cell body. Many axons are insulated by a fatty myelin sheath, which is interrupted at intervals by regions known as the nodes of Ranvier. The terminal branches of some axons make contact with as many as a thousand other neurons. The presynaptic terminals of the neuron in the diagram are shown synapsing on the cell bodies of a few postsynaptic neurons. (*Left,* originally appeared in S. Ramón y Cajal, *Histologie du système nerveux de l'homme et des vertèbres,* 2 vols. [Paris: Maloine, 1909, 1911], from J. De Felipe and E. G. Jones, *Cajal on the Cerebral Cortex* [New York: Oxford University Press, 1988]; *right,* after Kandel, Schwartz, and Jessell, *Principles of Neural Science.*)

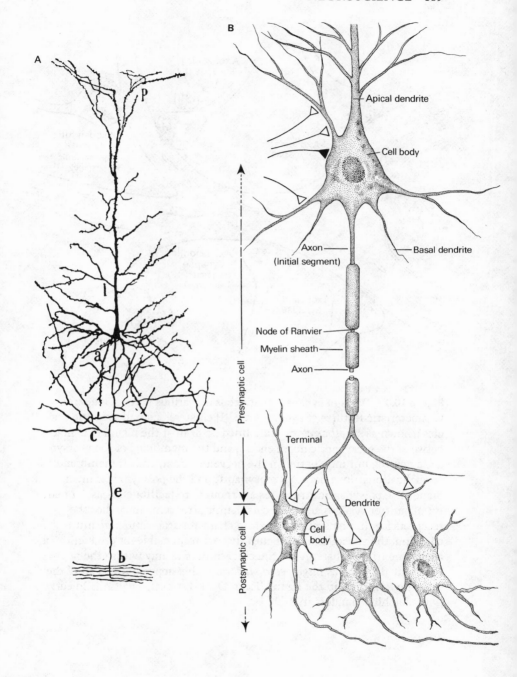

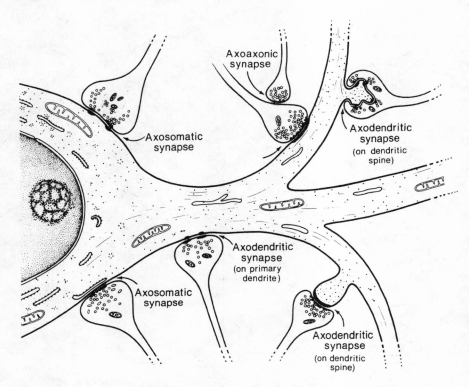

Figure 10.3 Types of synapses that occur on various parts of a neuron. Characteristic features of synapses are: (1) clusters of small synaptic vesicles (shown as *small circles*), some lined up against the region of contact between the presynaptic nerve ending and the membrane of the postsynaptic neuron, but mostly free in the nerve terminal; and (2) membrane specializations along both the presynaptic and the postsynaptic membrane. The presynaptic structures are thought to facilitate release of neurotransmitter vesicles and the postsynaptic structures to anchor the receptors for the transmitters. A third characteristic feature are mitochondria, the subcellular organelles that are responsible for production of adenosine triphosphate (ATP), the chief form of energy within the neuron. These are depicted as *large ovoid structures* that appear within all of the synaptic terminals in the figure. (From D. W. Fawcett, *The Cell*, 2d ed. [Philadelphia: Saunders, 1981].)

ceptors for all transmitters are made of protein molecules. Their structures and how they are regulated are being deciphered by the tools recently made available by the genetic revolution, and this information is illuminating the neurosciences. Many of the actions of nerve cells are now understood at the molecular level. To illustrate these advances, I shall discuss some examples of these actions in an order that follows the course of nerve action as just described. These actions range from how the nerve impulse is initiated to how neurotransmitters produce their effects when they bind to receptor proteins on postsynaptic target cells.

How Neurons Conduct Information

All living cells maintain an electrical charge difference across their external membrane. This separation of charge results in an electrical potential, with the inside negative (a measure of the energy available to drive ions through the membrane); in neurons it is about -60 mV. Positively charged sodium ions are normally in high concentration in the fluid that surrounds cells, and positively charged potassium ions are concentrated inside. In neurons, as well as in other excitable cells (e.g., as in muscle), certain stimuli result in a decrease in the potential across the membrane. This decrease, which is called depolarization, occurs because the neuron's external membrane suddenly becomes permeable to sodium ions, making the inside less negative. Thus, in these cells current actually flows.

The transient decrease in potential, called the action potential, is generated when the membrane of the axon is altered so that it becomes permeable to sodium ions. The mechanism for this change was worked out in the 1950s by Allen Hodgkin and Andrew Huxley, who used microelectrodes and biophysical modeling to suggest that a particular kind of ion channel, called *voltage-gated*, might be involved.[16] This kind of channel is now known to be a protein with a pore that opens when the electrical potential across the membrane decreases. When the sodium channel opens, a relatively large number of sodium ions flow into the cytoplasm of the axon from the extracellular fluid, thereby making the inside less negative with respect to the outside. But depolarization in a localized region of the membrane alone is insufficient to cause propagation. Another property of the axon is necessary for propagation: the depolarization is regenerative because neighboring channels open in response to the change in voltage. The depolarization thus sweeps down the axon in what is called the *nerve impulse*.

DNA technology has begun to provide a molecular explanation for the mechanism of voltage-gating. On this basis of structural information about three types of voltage-gated ion channel molecules: those for sodium, potassium, and calcium. This information was first obtained for the sodium channels, whose protein could be isolated because it binds a specific radioactively labeled toxin from scorpion venom. Although the biophysical theory of the action potential was well advanced by 1960, the biochemistry of the sodium channel—the structure of the molecule responsible—had to await cDNA cloning methods.[17]

Neurons also have channels for transporting potassium ions. These channels, concentrated in the membranes of nerve cell bodies and at the terminals, allow potassium ions to flow out of the cell when the membrane has been depolarized, thus helping to restore the membrane potential to its resting value. There are several varieties of potassium channels, which can be distinguished by the extent of their voltage-gating, their dependence on calcium ion, and their response to inhibitors.

Although potassium channels are the most common type of ion channel in nature, being widely distributed among plants as well as animals, they are present in low concentration and hence have been difficult to isolate. About a decade ago electrophysiological evidence suggested that *Shaker*, a mutant of the fruit fly (*Drosophila*) that twitches under certain conditions, is defective in the functioning of a potassium channel. In the first decade of this century, Thomas Hunt Morgan initiated genetic research with *Drosophila* because flies are easy to maintain in the laboratory and have a short generation time. During the intervening years the loci of many genes on fly chromosomes have been mapped and many useful techniques for manipulating those genes developed. With the help of the *Shaker* mutation, *Drosophila* genes for voltage-gated potassium channels have recently been cloned.[18]

Potassium channel proteins are quite similar in overall structure to sodium and calcium channels, which also have been cloned and isolated biochemically in other animals. Examination of the genes and corresponding amino acid sequences of these three kinds of ion channel molecules indicate that they are all closely related or belong to the same family of proteins. As implied by the word *family*, these contemporary channel molecules are thought to derive from a common ancestral gene that duplicated and diverged during evolution into the contemporary family of voltage-gated ion channels.[18,19]

Knowledge of the structure of the ion channel proteins inferred from DNA cloning experiments led to an explanation of voltage-gating. Ordinarily the parts of a protein that span a cell's membrane contain only nonpolar amino acids, which have affinity for the fatty substances (lipids) of the membrane, while the parts with charged amino acids come into contact with the aqueous solution on either side. Segments of sodium, potassium, and calcium channel protein molecules, however, are unusual because they have sequences containing polar amino acids *within* the membrane. These charged units are thought to operate as voltage sensors, that is, they move in response to depolarization of the membrane, thus changing the shape of the molecule and causing the channel to open.

The second function that has been explained in the sequence of nerve action is the molecular basis of insulation. For many behaviors rapid conduction of the nerve impulse is valuable. An obvious instance is response to danger. In vertebrates nature increases the speed of this response by insulating the axon with *myelin*, a sheath composed of many layers of electrically resistant membrane. The myelin sheath is formed during development of the nervous system, when specialized support cells called *glia* surround the axon to form a double-membrane structure, which wraps around the axon many times like a jelly roll. In mature myelin, the glial cytoplasm is squeezed out from between the fatty membrane layers, which condense to form the compact myelin sheath.

Patchy loss of myelin is the main defect in a relatively common human disease, multiple sclerosis. This loss interferes with impulse conduction, and therefore with both sensory perception and motor co-ordination. An experimental model for multiple sclerosis, called experimental allergic encephalomyelitis, results from the immune response when characteristic basic proteins of myelin are injected into animals.

Another important animal model is the *Shiverer (shi)* mutant of mice, in which much of the gene for myelin basic protein has been deleted. In mice that are homozygous *(shi/shi)*, myelination in the central nervous system is reduced to less than 10 percent, and the mice convulse and die young. When the wild-type gene is microinjected into fertilized *shi/shi* eggs the resulting transgenic mice express the wild-type gene at the correct stage of development and show improved myelination. Although these mice still have occasional tremors, they do not convulse or suffer early death.[20] This experiment is a promising example of partially successful gene therapy for a disease of the nervous system.

How Neurons Send Information to Other Cells

Progress has been prodigious in understanding receptors for neurotransmitters. For a long time, receptors were only a formal concept, as were genes before James D. Watson and Francis Crick described the structure of DNA.[21] For example, at the turn of the century Paul Ehrlich first proposed that both drugs and endogenous regulators like hormones must work at specific sites within particular cells.[22] Until the 1960s it was still uncertain whether receptors are proteins. Today many receptors for neurotransmitters have been isolated and fully characterized, and their role in memory is being elucidated.

Ionic Signaling

As a prime example, the peripheral receptor for the small-molecule neurotransmitter acetylcholine is perhaps the most completely characterized vertebrate membrane protein. This receptor, located at synapses on muscle, consists of five separate protein subunits, two of which are alike. Each subunit is firmly anchored in the muscle's membrane by four membrane-spanning segments. The five subunits aggregate to form a barrel-shaped channel through the muscle's membrane. This postsynaptic channel is closed until two molecules of acetylcholine released from the presynaptic neuron bind to a specific site on each of the two identical subunits in a region exposed at the cell surface. Binding causes the segments of the proteins that span the membrane to change their shape, allowing sodium and other ions to pass through the membrane and thus triggering the muscle to contract.

These ideas about how the acetylcholine receptor operates are based on knowledge of its protein structure obtained through the use of recombinant DNA technology. With this tool, the complete amino acid sequence of each of the receptor's four different subunits was determined. Detailed molecular models of the structure of the ion channel in the membrane, suggested by these sequences, are now being checked by X-ray crystallography.[23]

The brain contains similar receptors for acetylcholine, as well as receptors for other neurotransmitters. The principal inhibitory receptor opens a chloride ion channel when it binds the small-molecule transmitter gamma-aminobutyric acid. Other inhibitory receptors respond to the transmitter glycine.[24] These receptors resemble the acetylcholine receptor in general structure.

Biochemical Signaling

In the 1970s these neurotransmitter-gated ion channels were thought to be the only kind of receptor involved in synaptic transmission. But the elaboration of the concept of signal transduction (the re-encoding of information across cell's membrane) led to another class of neurotransmitter receptors.[25] Unlike ion channel receptors, in which binding of transmitter and opening of a channel are properties of the same protein molecule, receptors for signal transduction, when stimulated by binding a transmitter substance, change their shape in a way that activates several distinct protein molecules within the cell. These act in a sequence (called a *cascade*) to produce a change in the biochemical state of the cell.

One of the first examples of signal transduction to have been studied is the response of the membrane-spanning protein rhodopsin to light. Through a series of biochemical reactions produced by the energy of a photon on rhodopsin, an external physical signal is recoded into another form, an electrical signal, within the rod cell of the retina. In this cell the transduced signal ultimately takes the form of electrical impulses traveling to the brain.

The intermediate steps in signal transduction were clarified in 1950 by the discovery of cyclic 3', 5'-adenosine monophosphate (cyclic AMP).[26] Neurotransmitters reacting with receptors on the cell surface activate a G-protein (so called because it binds guanosine nucleotides), which in turn activates an enzyme called *adenyl cyclase*, to produce many molecules of cyclic AMP within the cell. This compound triggers a cascade of biochemical effects (Figure 10.4). Cyclic AMP and several substances produced through other signal transduction processes are called *second messengers* (the first messenger being the neurotransmitter or light).

Because they are synthesized in relatively large amounts, these second messengers in turn can amplify the input signal by modifying the activities of many enzymes and proteins within the cell. For example, cyclic AMP activates enzymes that catalyze the transfer of negatively charged phosphoryl groups to a large number of proteins within the cell. Phosphorylation can either activate or inhibit these substrate proteins. When phosphorylated, some of these proteins affect ion channels and the cytoskeletal protein fibers that move synaptic vesicles to sites of release. Cyclic AMP and other second-messenger reaction cascades can thus modulate synaptic transmission both indirectly, by influencing ion channels, and directly, by regulating the availability of synaptic vesicles for release.

Several second-messenger cascades are known to modulate synaptic release or to change the properties of receptor molecules themselves. In addition to cyclic AMP, there are other second messengers, some derived from the structural phospholipids in the surface membrane of the neuron.[27]

With extraordinary speed, recombinant DNA technology, coupled with modern biochemical methods, has disclosed that all of these modulating receptors are similar in structure and also belong to a single gene family.[28] Moreover, they are all related to the photopigment protein of the eye, rhodopsin. Like rhodopsin, each operates through the cell's external membrane to associate with a G-protein on the membrane's cytoplasmic side. This association in turn produces a change in another protein, such as the enzyme that synthesizes cyclic AMP.[29] (This change can be either stimulatory or inhibitory, depending on the particular receptor and on the particular G-protein it binds.) Cyclic AMP then activates still other enzymes that phosphorylate many of the cell's proteins. *Second-messenger cascade* is thus an apt name for this kind of synaptic action.

Figure 10.4   Two families of receptors that mediate synaptic transmission: neurotransmitter-gated ion channel receptors, and receptors that mediate signal transduction. *Left,* Neurotransmitter-gated ion channel receptors are illustrated by the nicotinic acetylcholine receptor that is present at the nerve-muscle synapse. This receptor is formed from five protein subunits that extend through the external membrane of the muscle. In upper left (closed), without the neurotransmitter acetylcholine (represented by a *black T*), ions do not pass through the receptor. In lower left (open), the motor neuron (the presynaptic cell) releases acetylcholine. Two molecules of the transmitter bind to the extracellular portion of the receptor, causing the molecule to change its shape and form an open pore through which sodium ions (Na^+) and potassium ions (K^+) can pass freely. This results in the rapid depolarization of the muscle cell. *Right,* Receptors that mediate signal transduction are illustrated by a receptor that can activate the enzyme adenyl cyclase. This kind of receptor is formed from a single polypeptide chain that spans the membrane seven times. In upper right (inactive), without a transmitter, the cyclase is inactive. In lower right (activated), a presynaptic neuron releases the proper neurotransmitter (represented by a *black triangle*), which binds to the receptor. The receptor then activates a G-protein, which stimulates adenyl cyclase to make cyclic AMP from ATP. Cyclic AMP acts as a second messenger and, among other effects, can influence ion channels to depolarize the postsynaptic cell. Thus, there is an important molecular difference between the kinds of synaptic transmission mediated by these two families of receptors. The effects of the first type are all a function of a single receptor protein molecule, while those of the second are produced by several distinct protein molecules that can be situated at some distance from one another. (After Kandel, Schwartz, and Jessell, *Principles of Neural Science.*)

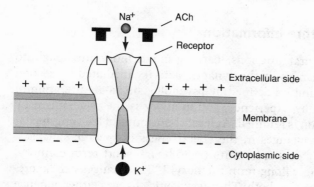

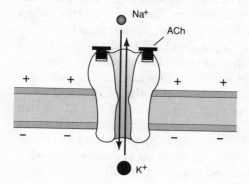

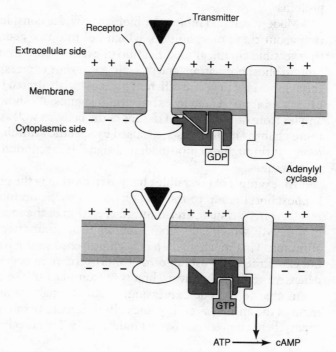

How Neurons Store Information

The higher mental functions, learning and memory, are thought to be produced by changes in the synaptic activity mediated by second messengers. Learning, as defined by psychologists, is an alteration of behavior produced by experience, and memory is the mechanism by which that alteration is retained. At least some forms of learning appear to be produced by changes in the effectiveness of specific synapses. These changes may persist for minutes to hours (short-term memory) or for weeks or longer (long-term memory).[30] The changes can be pre-synaptic, postsynaptic, or both. The presynaptic changes either enhance or diminish the release of neurotransmitter. The postsynaptic changes affect the functional properties of receptors for neurotransmitters.

Neurotransmitter release is controlled both by the availability of vesicles for release and by the flow of calcium ions into the terminal. The entry of calcium ions can be modified, either directly or indirectly, by changing the flow of other ions through still other ion channels.[31] In addition to the macromolecules of the ion channels, the fibers that make up the cytoskeleton and the receptors for neurotransmitters in the postsynaptic membrane are also proteins. Thus memory and learning are reduced to a tractable biological problem of determining what molecular mechanisms regulate the function of all these cellular proteins.

Modern cell and molecular biology provides considerable information about these mechanisms, which fall into two general categories: (1) reversible chemical modification of existing protein molecules, like protein phosphorylation, and (2) changes in gene expression resulting in the synthesis of new and different proteins with altered functions. Both categories are triggered by extracellular signals. As shown by many examples from other cells and tissues of the body, as well as from bacteria, these changes in proteins are caused by specific stimuli from the environment through receptor-mediated signal transduction across the cell membrane.

An example of reversible changes in protein is the phosphorylation (a phosphoryl group to the side chain of specific amino acid residues) resulting from transmitter-induced elevation of the second messenger, cyclic AMP, which lasts from minutes to an hour after an appropriate stimulus. This mechanism has been proposed as the basis of a simple but experimentally useful form of short-term memory: sensitization of defensive reflexes in invertebrates.[30] Examples of the second mechanism, change in gene expression, occur throughout normal development of the animal as embryonic cells differentiate into specialized cell types; these changes can persist indefinitely. This mechanism is widely thought to underlie long-term memory.[32] Although the neurons at which these changes take place can be identified in simpler invertebrate animals, localizing them in higher animals remains an enormous challenge.

In summary, many of the macromolecules important to nerve function have been isolated or cloned and their amino acid sequences determined. More significant, a genetic relationship has been established between molecules of similar function. For example, ion channels and receptors that form ion channels when they bind neurotransmitter are related, and they have probably evolved from a common ancestral ion channel. The receptors that activate second-messenger cascades all appear to belong to another gene family related to rhodopsin,[25,28] and the transducing guanosine nucleotide-binding proteins that they activate, which operate by moving macromolecules short distances across each other, belong to still another gene family.[29] Molecular studies permit synaptic transmission to be arranged into two categories. In one, neurotransmitters activate receptors that serve as ion channels. In the other, transmitters influence the biochemical properties of the target cell through signal transduction mechanisms that produce second messengers. These two forms of synaptic transmission are governed by a few distinct gene families. While new family members undoubtedly remain to be identified and analyzed, the emerging pattern is clear and simple.

Integrative Aspects of the Nervous System

So far, neural function has been described here primarily from the point of view of the single nerve cell and its molecular components. How the hundred billion neurons in the human brain are connected to one another is obviously also important. Molecular and cell biology is likely to have less impact on this work. Since the days of Penfield, much of the body's sensory system has been mapped to parts of the central nervous system. Internal representation of the world begins at the spinal cord, the first part of the central nervous system to receive sensory input from the periphery. This input is then relayed sequentially to ever-higher parts of the brain and finally to the cerebral cortex, which is the most recent and complex part of the brain to evolve.

Extraordinary advances have been made in localizing vision, in defining the organization of the brain centers involved in producing visual images, and in understanding how these perceptual centers develop in the growing animal. Similarly, knowledge about the organization of other sensory modalities, as well as motor function, has also made enormous strides.[12] Furthermore, analysis of the behavior of neural networks, both biological and artificially constructed by computers, is providing valuable insight into how ensembles of neurons operate.[33]

Methods for locating function can be greatly enhanced by techniques developed during the genetic revolution. For example, specific neurons can now be located more easily by procedures that make use of recombinant DNA. Thus, in some neurons involved in pain the various neuropeptide transmitter substances that modulate pain, such as opioid peptides (the endogenous opiates), are formed by transcription of messenger RNA translated from their corresponding genes. Radioactively labeled RNA or DNA sequences related to this messenger RNA can form complexes with it (called hybrids), and these can be detected within the cells in microscopic tissue sections by photographic emulsions that are sensitive to radioactivity, in a procedure known as *in situ hybridization*. Hence pain-specific nucleic acid sequences can be used as probes to identify pain cells.

Furthermore, the powerful, essentially noninvasive techniques of computerized isotopic imaging, such as positron-emission tomography (PET) scanning, now make it possible for clinical scientists to see regional changes in biochemical activity as they actually happen in the living brain. In this procedure a harmless, radioactively labeled substance is injected into a human subject, accumulates in active brain cells, and is located by radiation detected outside the body.[34]

Conclusion

The impact of the genetic revolution has two aspects; scientific and social. Scientifically, cell and molecular biology has had an almost unqualified triumph because it provides the most fruitful techniques for understanding the nervous system. Great progress is already evident in understanding how neurons are constructed at a molecular level and how they conduct, transfer, and store information. As an analytical method for isolating and determining the structures of functionally important molecules that are present only in tiny amounts such as potassium channels, recombinant DNA technology is accelerating in importance. The amount of functional information that can be derived from knowing the structure of proteins is astonishing.

In addition to protein structure, analysis of gene structure and expression, as well as of the biology of cells other than neurons, has brought within reach the complex problems of how the brain develops. Peptide or protein growth factors, some related to long-familiar hormones such as insulin, are elaborated by some neurons or by glial cells. These factors, which are required to permit development in the maturing brain and to promote regeneration in the pheripheral nervous system, are being identified and characterized.[35] It seems likely that these factors will be used therapeutically to promote regeneration of damaged adult brain tissue.

Application of the new methods to the problems of developmental biology can be expected to have an even more general impact, in two ways: understanding the mechanisms that turn various genes on and off in specific neurons at specified times in an orderly sequence, and identifying the macromolecules whose interactions govern the shapes of neurons and the contacts they make with other cells. These advances should play a major role in revealing how specific connections are formed in the developing brain, and also in approaching learning and memory as aspects of neuronal growth or late development.

Neuroscientists and philosophers of science frequently question whether understanding molecules can ever lead to a full explanation of brain mechanisms. Understanding how the cellular units of the brain operate thus would be only a trivial part of understanding the complex functioning of the brain. This position is captured in the anti-reductionist argument that an understanding of the parts of a radio can never alone explain how the instrument works. Proponents of this view do not expect molecular biology to have important scientific impact on the most challenging problems of neuroscience.[36]

This pessimism about reductionism is philosophically naive but widespread. Educated people and even physicians, upon learning about the benefits of neuropharmacology in the treatment of depression or schizophrenia, or about the modulations of synaptic transmission that underlie learning and memory, often ask "Is *that* all there is?" Since the beginning of the twentieth century, science has had the materialistic metaphysic that at one level matter and energy *are* all there is. Obviously, an understanding of neurons as the basic units of brain function is *not* "all there is," just as knowledge about atoms does not provide all the information needed to understand chemistry. Nevertheless, just as the atomic table holds the key to understanding chemistry, knowledge about the neuron is needed to explain how populations of neurons behave. Science's answer to the radio analogy has always been clear and definite: materialism implies that the whole *is* the sum of its parts, interacting in some organized and comprehensible manner.

Whatever the scientific benefits of molecular neurobiology, there are also disadvantages. The deepest questions to be answered by the neural sciences derive not so much from science itself, in the strict sense of the specifically Western activity that began in the seventeenth century,[37] as from natural history, philosophical speculation, and clinical practice that, since antiquity have provided the impetus for inquiry. Questions about the nature of the mind were formulated to understand consciousness, a fundamental mystery of life.[38] Other approaches and ways of experiencing mysteries and of formulating questions—through simple observation, dynamic psychology, ethology, historical and comparative social studies—may be discouraged precisely because molecular neuroscience is so successful in what it can do.[39] The important scientific question may be not, Is that all there is? but rather, Will we miss something by focusing too much on the molecular biology of the neuron?

In addition to its scientific aspect, the genetic revolution in neuroscience has a social impact. The experiments of Jacques Loeb and his enthusiasm for the control of life phenomena, especially reproductive processes,[40] provoked Aldous Huxley to write *Brave New World* (1932). There is likely to be a similar public response to molecular neurobiology, where the life phenomenon under consideration is the human essence: the mind.

Certainly the question Is that all there is? has a different implication for the layperson than for the scientist. Even to the secular layperson, mind, self, and personality are the last parts to be surrendered willingly to "natural law." Accordingly, as the genetic revolution invades this area it may result in immense disappointment and resentment, and may further the current popular trend toward irrationality and fundamentalism.

The genetic revolution is an extension of secular ideas and technologies that have dominated Western thought at least since the middle of the nineteenth century.[41] These have been congenial to modern industrial society because of their promise of progress, improvement, and higher standards of living. To some degree, this promise has been kept. But as some of the early proponents of Darwinism predicted,[42] these ideas and technologies have not been able to guarantee peace, happiness, and social well-being, because science and technology do not supply the values that are essential to underwrite these goals. If we expect much more in this area from our increasing understanding of the brain, we are likely to be disappointed.

While most people have little trouble accepting the idea of the heart as an organ whose function can be understood completely in terms of physics and chemistry, they may be reluctant to approach the brain in these terms. Molecular neuroscience may divert attention from the essential problem of consciousness, and it certainly involves relinquishing the soul as an entity distinct from a material base. Yielding the mind to "natural law" raises existential questions that will surely generate serious tensions.

Notes

I thank Patricia Churchland, Anne Harrington, Evelyn Fox Keller, and Edward Manier for their helpful discussions on topics relating to this chapter.

1. J. Loeb, "Experimental Study of the Influence of Environment on Animals," in *Darwin and Modern Science,* ed. A. C. Seward (New York: G. P. Putnam's, 1909, pp. 247–70.

2. R. M. Young, "The Impact of Darwin on Conventional Thought," in *Darwin's Metaphor, Nature's Place in Victorian Culture* (Cambridge: Cambridge University Press, 1985).

3. J. G. Spurzheim, *The Physiognomical System of Drs. Gall and Spurzheim,* 2d rev. ed. (London, 1815), pp. 15, 57–60, 164.

4. The number of organs recognized by phrenologists increased during the nineteenth century. Spurzheim (ibid.) lists a smaller number than do the later practitioners of cranioscopy. An extensive list of the phrenological faculties can be found in the innumerable nineteenth-century publications of O. S. and L. N. Fowler, New York, which later became the S. R. Wells and Co. and, still later, Fowler and Wells Co. in New York, and L. N. Fowler and Co. in London. See also n. 6.

5. R. Cooter, *The Cultural Meaning of Popular Science: Phrenology and the Organization of Consent in Nineteenth-Century Britain* (Cambridge: Cambridge University Press, 1984).

6. M. B. Stern, *Heads and Headlines: The Phrenological Fowlers* (Norman: University of Oklahoma Press, 1971).

7. W. Penfield and T. Rasmussen, *The Cerebral Cortex of Man: A Clinical Study of Localization of Function* (New York: Macmillan, 1950).

8. The scientific, intellectual, and social history of the localizationist point of view and the extraordinarily strong opposition to it have been described in three outstanding books. In addition to Cooter, *Cultural Meaning,* see R. M. Young, *Mind, Brain, and Adaptation in the Nineteenth Century* (Oxford: Clarendon Press, 1970), and A. Harrington, *Medicine, Mind, and the Double Brain: A Study in Nineteenth-Century Thought* (Princeton: Princeton University Press, 1987).

9. P. J. Pauly, *Controlling Life: Jacques Loeb and the Engineering Ideal in Biology* (New York: Oxford University Press, 1987); and W. James, *The Principles of Psychology,* 2 vols. (New York, 1890) 1: 30–80.

10. K. S. Lashley, "In Search of the Engram," *Symp. Soc. Exp. Biol.* 4 (1950): 454–82.

11. M. S. Gazzaniga, J. E. Bogen, and R. W. Sperry, "Some Functional Effects of Sectioning the Cerebral Commisure in Man," *Proc. Natl. Acad. Sci. USA* 48 (1962): 1765–69; M. S. Gazzaniga, *Handbook of Cognitive Neuroscience* (New York: Plenum, 1984); N. Geschwind and A. M. Galaburda, *Cerebral Lateralization, Biological Mechanisms, Associations, and Pathology* (Cambridge, Mass.: MIT Press, 1987).

12. *Somatosensory:* V. B. Mountcastle and E. Henneman, "The Represen-

tation of Tactile Sensibility in the Thalamus of the Monkey," *J. Comp. Neurol.* 97 (1952): 409–36.

Pain: E. J. Simon and J. M. Hiller, "Opioid Peptides and Opioid Receptors," in *Basic Neurochemistry,* ed. G. Siegel, B. Agranoff, R. W. Albers, and P. Molinoff, 4th ed. (New York: Raven Press, 1989), pp. 271–85.

Visual: D. H. Hubel and T. N. Wiesel, "Ferrier Lecture: Functional Architecture of Macaque Monkey Visual Cortex," *Proc. R. Soc. Lond. [Biol.]* 198 (1977): 1–59; M. S. Livingstone and D. H. Hubel, "Connections between Layer 413 of Area 17 and the Thick Cytochrome Oxidase Stripes of Area 18 in the Squirrel Monkey," *J. Neurosci.* 7 (1987): 3371–77; D. H. Hubel and M. S. Livingstone, "Segregation of Form, Color, and Stereopsis in Primate Area 18," *J. Neurosci.* 7 (1987): 3378–415; M. S. Livingstone and D. H. Hubel, "Psychophysical Evidence for Separate Channels for the Perception of Form, Color, Movement, and Depth," *J. Neurosci.* 7 (1987): 3416–18.

Cerebellar Function: M. Ito, *The Cerebellar and Neural Control* (New York: Raven Press, 1984).

Higher functions: N. Geschwind and A. M. Galaburda, eds., *Cerebral Dominance: The Biological Foundations* (Cambridge, Mass.: Harvard University Press, 1984); J. E. LeDoux and W. Hirst, eds., *Mind and Brain: Dialogues in Cognitive Neuroscience* (Cambridge: Cambridge University Press, 1986); G. A. Ojemann and C. Mateer, "Human Language Cortex: Localization of Memory, Syntax, and Sequential Motor-Phoneme Identification Systems," *Science* 205 (1979): 1401–3.

13. E. R. Kandel, J. H. Schwartz, and T. Jessell, *Principles of Neural Science,* 3d ed. (New York: Elsevier, 1991); J. R. Cooper, F. E. Bloom, and R. H. Roth, *The Biochemical Basis of Neuropharmacology,* 6th ed. (New York: Oxford University Press, 1991).

14. P. Fatt and B. Katz, "An Analysis of the End-plate Potential Recorded with an Intra-cellular Electrode," *J. Physiol. Lond.* 115 (1951): 320–70; J. Del Castillo and B. Katz, "La base 'quantale' de la transmission neuro-musculaire," *Microphysiologie comparée des éléments excitables, Colloq. Int. Cent. Nat. Rech. Sci.* 67 (1957): 245–58.

15. S. L. Palay, "The Morphology of Synapses in the Central Nervous System," *Exp. Cell. Res.,* Suppl., 5 (1958): 275–93; V. P. Whittaker, "The Isolation and Characterization of Acetylcholine-containing Particles from Brain," *Biochem. J.* 72 (1959): 694–706; E. DeRobertis, "Submicroscopic Morphology and Formation of the Synapse," *Exp. Cell Res.,* Suppl., 5 (1958): 347–69; E. DeRobertis and H. S. Bennett, "Some Features of the Submicroscopic Morphology of Synapses in Frog and Earthworm," *J. Biophys. Biochem. Cytol.* 1 (1955): 47–58; V. P. Whittaker, I. A. Michaelson, and R. J. A. Kirkland, "The Separation of Synaptic Vesicles from Nerve-ending Particles ('Synaptosomes')" *Biochem. J.* 90 (1964): 293–303.

16. A. L. Hodgkin and A. F. Huxley, "A Quantitative Description of Membrane Current and Its Application to Conduction and Excitation in Nerve," *J. Physiol. Lond.* 117 (1952): 500–544.

17. W. A. Catterall, "The Molecular Basis of Neuronal Excitability," *Sci-*

ence 223 (1984): 653–61; M. Noda, S. Shimuzu, T. Tanabe, T. Takai, T. Kayano, T. Ikeda, H. Takahashi, H. Nakayama, Y. Kanaoka, N. Minamino, K. Kangawa, H. Matsuo, M. Raftery, T. Hirose, S. Inayama, H. Hayashida, T. Miyata, and S. Numa, "Primary Structure of *Electrophorus Electricus* Sodium Channel Deduced from cDNA Sequence," *Nature* 312 (1984): 121–27; R. L. Barchi, "Probing the Molecular Structure of the Voltage-dependent Sodium Channel," *Annu. Rev. Neurosci.* 11 (1988): 455–95.

18.   L. Y. Jan and Y. N. Jan, "Voltage-sensitive Ion Channels," *Cell* 56 (1989): 13–25.

19.   B. Hille, *Ionic Channels of Excitable Membranes* (Sunderland, Mass.: Sinauer, 1984), pp. 371–83.

20.   C. Readhead, B. Popko, N. Takahasi, H. D. Shine, R. A. Saavedra, R. L. Sidman, and L. Hood, "Expression of Myelin Basic Protein Gene in Transgenic *Shiverer* Mice: Correction of the Dysmyelinating Phenotype," *Cell* 48 (1987): 703–12; G. Lemke, "Unwrapping the Genes of Myelin," *Neuron* 1 (1988): 535–43.

21.   J. D. Watson and F. H. C. Crick, "The Structure of DNA," *Cold Spring Harbor Symp. Quant. Biol.* 18 (1953): 123–31.

22.   P. Ehrlich, "On Immunity, with Special Reference to Cell Life," Croonian Lecture, *Proc. R. Soc. Lond.* 66 (1900): 424–48.

23   R. M. Stroud and J. Finer-Moore, "Acetylcholine Receptor Structure, Function, and Evolution," *Annu. Rev. Cell. Biol.* 1 (1985): 317–51.

24.   G. Grenningloh, A. Rienitz, B. Schmitt, C. Methfessel, M. Zensen, K. Beyreuther, E. D. Gundelfinger, and H. Betz, "The Strychnine-binding Subunit of the Glycine Receptor Shows Homology with the Nicotinic Acetylcholine Receptor," *Nature* 328 (1987): 215–20; P. R. Schofield, M. G. Darlison, N. Fugita, D. R. Burt, R. A. Stephenson, H. Rodriguez, L. M. Rhee, J. Ramachandrah, V. Reale, T. A. Glencorse, P. H. Seeburg, and E. A. Barnard, "Sequence and Functional Expression of the GABA Receptor Shows a Ligand-gated Receptor Super-family," *Nature* 328 (1987): 221–27.

25.   Molecular biology of signal transduction. *Cold Spring Harbor Symp. Quant. Biol.* 53 (1988).

26.   Reviewed by E. W. Sutherland and T. W. Rall in "The Relation of Adenosine 3', 5'-phosphate and Phosphorylase to the Actions of Catecholamines and Other Hormones," *Pharmacol. Rev.* 12 (1960): 265.

27.   E. J. Nestler and P. Greengard, *Protein Phosphorylation in the Nervous System* (New York: John Wiley, 1984); Y. Nishizuka, "The Molecular Heterogeneity of Protein Kinase C and Its Implications for Cellular Regulation," *Nature* 334 (1988): 661–64; M. J. Berridge, "Inositol Trisphosphate and Diacylglycerol: Two Interacting Second Messengers," *Annu. Rev. Biochem.* 56 (1987): 159-93; E. Shapiro, D. Piomelli, S. Feinmark, S. Vogel, G. Chin, and J. H. Schwartz, "The Role of Arachnidonic Acid Metabolites in Signal Transduction in an Identified Neural Network Mediating Presynaptic Inhibition in *Aplysia*," *Cold Spring Harbor Sym. Quant. Biol.* 53 (1988): 425–33. For summary, see L. W. Role and J. H. Schwartz, "Crosstalk between Signal Transduction Systems," *Trends in Neurosci.* 12 (1989): centerfold.

28.   B. F. O'Dowd, R. J. Lefkowitz, and M. G. Caron, "Structure of the Ad-

renergic and Related Receptors," *Annu. Rev. Neurosci.* 12 (1989): 67-83.

29. A. G. Gilman, "G Proteins: Transducers of Receptor-generated Signals," *Annu. Rev. Biochem.* 56 (1987): 615-49.

30. E. R. Kandel and J. H. Schwartz, "Molecular Biology of Learning: Modulation of Transmitter Release," *Science* 218 (1982): 433-43; R. G. M. Morris, E. R. Kandel, and L. R. Squire, "The Neuroscience of Learning and Memory: Cells, Neural Circuits and Memory," *Trends in Neurosci.* 11 (1988): 125–27; *Learning/Memory: A Special Issue, Trends in Neurosci.* 11 (1988) (entire issue); Y. Dudai, *The Neurobiology of Memory: Concepts, Findings, Trends* (New York: Oxford University Press, 1989); L. R. Squire, *Memory and Brain* (New York: Oxford University Press, 1987).

31. L. K. Kaczmarek and I. B. Levitan, *Neuromodulation: The Biochemical Control of Neuronal Excitability* (New York: Oxford University Press, 1987).

32. P. Goelet, V. F. Castellucci, S. Schacher, and E. R. Kandel, "The Long and Short of Long-term Memory: A Molecular Framework," *Nature* 322 (1986): 419–22.

33. A. I. Selverston, *Model Neural Networks and Behavior* (New York: Plenum Publishing, 1985); S. J. Hanson and C. R. Olson, eds, *Connectionist Modeling and Brain Function: The Developing Interface* (Cambridge, Mass.: MIT Press, 1990); D. E. Rumelhart and J. L. McClelland, *Parallel Distributed Processing: Explorations in the Microstructure of Cognition,* vol. I, *Foundations* (Cambridge, Mass.: MIT Press, 1986); E. L. Schwartz, ed., *Computational Neuroscience* (Cambridge, Mass.: MIT Press, 1990); T. J. Sejnowski, A. Koch, and P. S. Churchland, "Computational Neuroscience," *Science* 241 (1988): 1299–1306.

34. G. L. Brownell, T. F. Budinger, P. C. Lauterbur, and P. L. McGeer, "Positron Tomography and Nuclear Magnetic Resonance Imaging," *Science* 215 (1982): 619–26; W. Oldendorf and W. Oldendorf, Jr., *Basics of Magnetic Resonance Imaging* (Boston: Martinus Nijhoff, 1988); M. E. Raichle, "Circulatory and Metabolic Correlates of Brain Function in Normal Humans," *Handbook of Physiology, Section 1. The Nervous System, Vol. 5. Higher Functions of the Brain, Part 2,* ed. F. Plum (Bethesda, Md.: American Physiological Society, 1987), pp. 643–74; L. Sokoloff, "Modeling Metabolic Processes in the Brain in Vivo," *Annu. Rev. Neurol.,* Suppl., 15 (1984): S1–S11.

35. M. Schwartz, A. Cohen, C. Stein-Izsak, and M. Belkin, "Dichotomy of the Glial Cell Response to Axonal Injury and Regeneration," *FASEB J.* 3 (1989): 2371–78.

36. See discussion in E. R. Kandel, "Neurobiology and Molecular Biology: The Second Encounter," *Cold Spring Harbor Symp. Quant. Biol.* 43 (1983): 891–908.

37. See, for example, H. Butterfield, *The Origins of Modern Science, 1300–1800,* rev. ed. (London: G. Bell and Sons, 1957); S. Tomlinson, *Cosmopolis, The Hidden Agenda of Modernity* (New York: Free Press, 1990).

38. It is difficult to provide a comprehensive list of references for this point. For antiquity, see B. Snell, *The Discovery of the Mind* (Cambridge, Mass.: Harvard University Press, 1953); E. R. Dodds, *The Greeks and the Irrational*

(Berkeley: University of California Press, 1959); G. E. R. Lloyd, ed. *The Sacred Disease*, in *The Hippocratic Writings* (Harmondsworth: Penguin, 1978), 237–51; R. E. Siegel, *Galen on Psychology, Psychopathology and Function and Diseases of the Nervous System: An Analysis of His Doctrines, Observations, and Experiments* (Basel: S. Karger, 1973).

39. Important examples of fruitful approaches other than molecular neuroscience are C. Darwin, *The Expression of the Emotions in Man and Animals* (London, 1872); S. Freud, *The Ego and the Id*, trans. Joan Riviere (London: Hogarth Press, 1927); D. O. Hebb, *The Organization of Behavior: A Neuropsychological Theory* (New York: John Wiley, 1949); N. Tinbergen, *The Study of Instincts in Animals* (Oxford: Clarendon Press, 1951); K. von Frisch, *Bees: Their Vision, Chemical Senses, and Language* (Ithaca: Cornell University Press, 1950).

40. Through S. Lewis, *Arrowsmith* (1925); see also G. W. Corner, *A History of the Rockefeller Institute 1901–1953: Origins and Growth* (New York: Rockefeller Institute Press, 1964).

41. See, for example, O. Chadwick, *The Secularization of the European Mind in the Nineteenth Century: The Gifford Lectures in the University of Edinburgh for 1973–74* (Cambridge: Cambridge University Press, 1975); M. Mandelbaum, *History, Man, and Reason: A Study in Nineteenth-Century Thought* (Baltimore: Johns Hopkins University Press, 1971); R. M. Young, "Darwinism *Is* Social," in *The Darwinian Heritage*, ed. D. Korn (Princeton: Princeton University Press, 1985), 609–38.

42. T. H. Huxley, *Evolution and Ethics and Other Essays* (New York, 1898); see G. Himmelfarb, "The Victorian Trinity: Religion, Science, Morality," in *Marriage and Morals among the Victorians and Other Essays* (New York: Alfred A. Knopf, 1986).

11

Evolution

Francisco J. Ayala

The publication in 1859 of *The Origin of Species* by Charles Darwin ushered in a new era in intellectual history. Darwin is deservedly given credit for the theory of biological evolution: he accumulated evidence demonstrating that organisms evolve and discovered the process by which they evolve. But the import of this discovery transcends science. It radically changed our conception of the universe and our place in it. The impact of Darwin's intellectual revolution has now been enhanced by a new revolution in biology: the advances of molecular genetics have given scientists the tools for reconstructing the complete evolutionary history of organisms, including humans.

Darwin's Revolution

In the sixteenth and seventeenth centuries, the discoveries of Copernicus, Kepler, Galileo, and Newton gradually showed that the earth is not the center of the universe, as had previously been thought, but a small planet rotating around an average star; the universe was recognized as immense in space and in time; the motions of the planets around the sun could be explained by the same simple laws that accounted for the motion of physical objects on our planet. These and other discoveries greatly expanded human knowledge. But the conceptual revolution was even more fundamental, bringing a realization that the universe obeys immanent laws that can account for natural phenomena. The workings of the universe were brought into the realm of science: explanation through natural laws. Physical phenomena could be reliably predicted whenever the causes were adequately known.

Darwin completed the Copernican revolution by drawing out for biology the ultimate conclusion from the notion of nature as a lawful system of matter in motion. The adaptations and diversity of organisms, the origin of novel and highly organized forms, even the origin of man himself could now be explained by an orderly process of change governed by natural laws.

Before Darwin, the origin of organisms and their marvelous adaptations were most frequently attributed to the design of an omniscient Creator. God had created the birds and bees, the fish and corals, the trees in the forest, and, best of all, man. God had given us eyes so that we might see, and He had provided fish with gills to breathe in water. Philosophers and theologians argued that the functional design of organisms evinced the existence of an all-wise Creator. In the thirteenth century St. Thomas Aquinas had used the argument from design as his "fifth way" to demonstrate the existence of God. In the nineteenth century the English theologian William Paley argued in his *Natural Theology* (1802) that the functional design of the human eye provided conclusive evidence of an all-wise Creator. It would be absurd to suppose, wrote Paley, that the human eye by mere chance "should have consisted, first, of a series of transparent lenses . . . secondly of a black cloth or canvas spread out behind these lenses so as to receive the image formed by pencils of light transmitted through them, and placed at the precise geometrical distance at which, and at which alone, a distinct image could be formed . . . thirdly of a large nerve communicating between this membrane and the brain." The *Bridgewater Treatises*, published between 1833 and 1840, were written by eminent scientists and philosophers to set forth "the Power, Wisdom, and Goodness of God as manifested in the Creation." For example, the structure and mechanisms of the human hand were cited as incontrovertible evidence that it had been designed by the same omniscient Power that had created the world.

The advances of physical science had thus fostered a split-personality conception of the universe that persisted well into the mid-nineteenth century. Scientific explanations, derived from natural laws, dominated the world of nonliving matter, on the earth as well as in the heavens. Supernatural explanations, depending on the unfathomable deeds of the Creator, accounted for the origin and configuration of living creatures— the most diversified, complex, and interesting realities of the world. It was Darwin's genius to resolve this conceptual schizophrenia.

The conundrum faced by Darwin can hardly be overestimated. The strength of the argument from design to demonstrate the role of the Creator is easily set forth. Wherever there is function or design we look for its author. A knife is made for cutting, and a clock is made to tell time; their functional designs have been contrived by a knifemaker and a watchmaker. Similarly, the structures, organs, and behaviors of living beings are directly organized to serve certain functions. Hence, the functional design of organisms and their features would seem to argue for the existence of a designer. Darwin, however, showed that the directive organization of living beings can be explained without resorting to a Creator or other external agent.

Darwin recognized that organisms are functionally organized. They are adapted to certain ways of life, and their parts are adapted to perform certain functions. Fish are adapted to live in water, kidneys are designed to regulate the composition of blood, the hand is made for grasping. Darwin accepted the fact of adaptation and then showed that the process of natural selection provided a natural explanation for that fact. He thereby brought the issue of the origin and adaptative organization of living beings into the realm of science.

Darwin's theory encountered opposition in religious circles not so much because he proposed an evolutionary origin for living things—a proposal that had been made many times before, even by Christian theologians—but because his mechanism, natural selection, excluded God from the explanation of the adaptation of organisms. The Roman Catholic Church's opposition to Galileo in the seventeenth century had been similarly motivated.

Darwin supported his argument for the evolution of species with two sorts of evidence. The first came from paleontology. The fossil remains of organisms that lived in the past provide definite clues to their phylogeny, or evolutionary history. Phylogenetic inferences may also be made from the comparative study of living organisms. The logical basis of these inferences is simple. Because evolution is by and large a gradual process, organisms sharing a recent common ancestor are likely to be more similar to each other than are organisms with a common ancestor only in the remote past; relative degrees of similarity are therefore used to infer recency of common descent.

Figure 11.1 Similar structures of the forelimbs of four animals, even though the limbs are used for very different purposes: writing, running, swimming, flying. (From T. Dobzhansky, F. J. Ayala, G. L. Stebbins, and J. W. Valentine, *Evolution* [San Francisco: W. H. Freeman, 1977].)

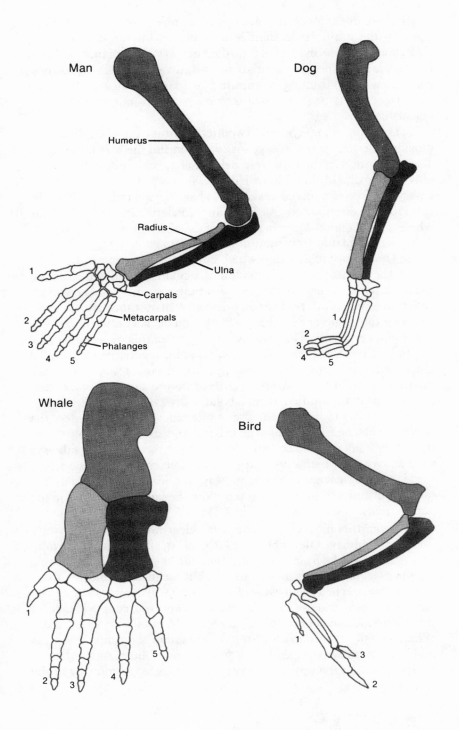

The argument for evolution based on comparative anatomy is illustrated by the similarity in the forelimbs of humans, dogs, whales, and birds, bone by bone, in spite of the different ways of life of these animals (see Figure 11.1). The same bone structures are used for such diverse purposes as writing, running, swimming, or flying, because these structures were inherited from a reptilian ancestor common to all of these organisms.[1]

Thus, until about the mid-twentieth century the disciplines that provided the strongest evidence for evolution and the best information about evolutionary history were paleontology and comparative anatomy, although additional knowledge was obtained from comparative embryology, comparative behavior, and biogeography (the study of the geographical distribution of organisms). Modern biology has, on the whole, corroborated Darwin's view of biological evolution and has added depth and infinitely more detail to our understanding of the processes. The science of genetics, which burst into existence with the advent of the twentieth century, made a particularly significant contribution. Darwin's theory of natural selection assumes that hereditary variation is pervasive. Yet Darwin knew no plausible mechanism to account for the origin of hereditary novelty and he had only tenuous evidence of its existence. Genetics eventually filled this major gap.

In addition, since the 1930s comparative biochemistry has provided much evidence for the unity of biology at a molecular level, underneath the infinite variety that we see in gross structure and function. Different organisms need different metabolic pathways (sequences of enzymatic reactions) to initiate the use of different foodstuffs to meet the two major biochemical needs of all cells: energy, and building blocks for the synthesis of specific cell constituents. But these diverse pathways soon converge on pathways that are the same throughout the living world, yielding energy in the same way, and leading to precisely the same set of amino acids in the proteins and the same nucleotides in the nucleic acids.

Though this biochemical unity had clear implications for evolutionary continuity, it never played a large role in efforts to defend evolution against attacks by creationists, or in efforts to trace the branching trees of past connections more precisely. But the fusion of biochemistry and genetics to yield the newest of all biological disciplines, molecular biology, has provided the most direct and reliable source of evidence for the reconstruction of evolutionary history. The possibility exists today of knowing the evolutionary history of any group of organisms with as much detail as wanted. Only the limitations of human or other resources stand in the way of reconstructing the grand panorama of the evolution of all life, from the lowly bacteria of 3 billion years ago to the microorganisms, animals, and plants of today.

The proteins and nucleic acids that are essential to the makeup of all organisms are "informational" macromolecules, which retain a record of their evolutionary history. The evolutionary information is contained in the linear sequence of their component elements, in much the same way as semantic information is contained in the sequence of letters of an English sentence. This evolutionary information is so detailed that it not only makes it possible to reconstruct the evolutionary tree, that is, the relationships of ancestry and branching in the history of organisms; it also opens up the possibility of timing the events in that history, even those that occurred in the remote past.

As a means to reconstruct evolutionary history, molecular biology has two notable advantages over comparative anatomy and the other classical disciplines. One is that the information is readily quantifiable: the number of component units that differ is established by comparing the sequence of the units in a given protein or gene from two different organisms. The other advantage is that quite different sorts of organisms can be compared. There is very little that comparative anatomy or embryology can say when the organisms being compared are as diverse as yeasts, pine trees, and human beings, but there are homologous macromolecules that can be compared in all three.

Molecular Evolution

Nucleic acids and proteins are linear molecules made up of units: four kinds of nucleotides in nucleic acids, twenty amino acids in proteins. Evolution typically occurs by the substitution of these units, one at a time. Comparison of a certain macromolecule in two species yields the number of differences, which is an indication of the recency of common ancestry, just as the number of miles separating two cars reflects how long they have been traveling in opposite directions. For example, the numbers of amino acid differences between three species—human, rhesus monkey, and horse—in their cytochrome c (a protein involved in electron transport in living cells) are one, twelve, and eleven, as shown in Table 11.1. The evolutionary history of these three species is known from classical evidence to be as represented in Figure 11.2. The data on amino acid differences (Table 11.1) confirm these lineages. It is evident from the number of differences that the last common ancestor of human and monkey lived much later than the last common ancestor of all three species. Two other phylogenies for the three species are formally possible (Figure 11.3), but they are ruled out because they require

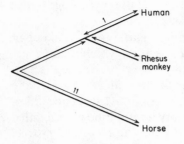

Figure 11.2 Number of amino acid substitutions in the phylogenetic history of the cytochrome c of three contemporary species: humans, rhesus monkeys, and horses.

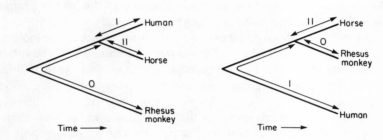

Figure 11.3 Two theoretically possible, but unlikely, phylogenies of humans, rhesus monkeys, and horses.

grossly disproportionate rates of evolution in different segments.

The data lead to an additional, more refined conclusion. Because there is one more difference between human and horse than between monkey and horse, that difference must be due to replacement of one amino acid by another in the human lineage rather than in the monkey lineage. This procedure can be extended to all the branches in a complex phylogeny. Thus, when only three species are compared, as in Figure 11.2, the data do not say how the eleven amino acid substitutions between the horse and the common ancestor of all three organisms, and between these and the common ancestor of the primates, are distributed between the two branches. This, however, can be resolved by comparison with a fourth species (say, the rattlesnake) that diverged from the other three before they diverged from each other. An interesting problem is that when the number of species in a phylogeny is greater than three, the calculation of the number of substitutions in each segment is "overdetermined": that is, there is more than one solution to the distribution of the substitutions among the lineages. Statistical methods (such as maximum likelihood or parsimony) are used for deciding which one is the "best" solution.

In one of the most dramatic early applications of molecular biology

Table 11.1 Number of amino acid substitutions between the cytochrome *c* molecules of humans, rhesus monkeys, and horses

	Human	Rhesus Monkey	Horse
Human	—	1	12
Rhesus monkey		—	11
Horse			—

Note: The cytochrome *c* in these organisms has 104 amino acids.

to the reconstruction of phylogeny, Walter Fitch and Emanuel Margoliash[2] calculated the number of substitutions necessary in the DNA to account for the amino acid differences in the cytochrome *c* molecules of twenty organisms (see Table 11.2). A phylogeny based on that data matrix is shown in Figure 11.4, which also gives the number of DNA (nucleotide) substitutions inferred for each branch. The phylogenetic relationships obtained with the molecular data correspond well, on the whole, with those determined from the fossil record and other sources. There are three conspicuous disagreements, however: chickens erroneously appear more closely related to penguins than to ducks and pigeons; the turtle, a reptile, appears closer to the birds than to the rattlesnake; humans and primates appear to diverge from the other mammals before the marsupial kangaroo separates from the nonprimate placentals. Despite these erroneous relationships, it is remarkable that the study of a single protein, and a small one at that, should yield a fairly accurate representation of the overall phylogeny of twenty organisms so diverse. The errors of detail can be corrected by reference to other proteins.

Cytochrome *c* molecules are slowly evolving proteins; that is, the rate of amino acid substitutions per unit of time is low. Therefore, organisms as different as humans, moths, and Neurospora molds have a large proportion of amino acids in their cytochrome *c* molecules in common. This evolutionary conservation makes possible the study of genetic differences among organisms that are only remotely related. However, this same conservation makes cytochrome *c* useless for determining evolutionary change in closely related organisms, such as different apes or chicken and duck, since these may have cytochrome *c* molecules that are completely or nearly identical, and the few differences that exist are subject to sampling errors, just as occurs when we toss a coin twice and get two heads even though the expectation is one head and one tail. Indeed, the primary structure of cytochrome *c* is identical in humans and chimpanzees, which diverged 5–10 million years ago, and it differs by only one amino acid between humans and rhesus

Table 11.2   Substitutions in the gene coding for cytochrome *c* in twenty organisms

Organism	1	2	3	4	5	6	7	8	9	10	11	12	13	14	15	16	17	18	19	20
1. Human	—	1	13	17	16	13	12	12	17	16	18	18	19	20	31	33	36	63	56	66
2. Monkey		—	12	16	15	12	11	13	16	15	17	17	18	21	32	32	35	62	57	65
3. Dog			—	10	8	4	6	7	12	12	14	14	13	30	29	24	28	64	61	66
4. Horse				—	1	5	11	11	16	16	16	17	16	32	27	24	33	64	60	68
5. Donkey					—	4	10	12	15	15	15	16	15	31	26	25	32	64	59	67
6. Pig						—	6	7	13	13	13	14	13	30	25	26	31	64	59	67
7. Rabbit							—	7	10	8	11	11	11	25	26	23	29	62	59	67
8. Kangaroo								—	14	14	15	13	14	30	27	26	31	66	58	68
9. Duck									—	3	3	3	7	24	26	25	29	61	62	66
10. Pigeon										—	4	4	8	24	27	26	30	59	62	66
11. Chicken											—	2	8	28	26	26	31	61	62	66
12. Penguin												—	8	28	27	28	30	62	61	65
13. Turtle													—	30	27	30	33	65	64	67
14. Rattlesnake														—	38	40	41	61	61	69
15. Tuna															—	34	41	72	66	69
16. Screwworm fly																—	16	58	63	65
17. Moth																	—	59	60	61
18. *Neurospora*																		—	57	61
19. *Saccharomyces*																			—	41
20. *Candida*																				—

Source: W. M. Fitch and E. Margoliash, "Construction of Phylogenetic Trees," *Science* 155 (1967): 279–84.

monkeys, whose most recent common ancestor lived 40–50 million years ago.

Fortunately, different proteins evolve at different rates (Figure 11.5). If a protein has to fit into an organized structure, such as a membrane or a complex of proteins with each other or with nucleic acids, much of its surface has to fit the neighboring molecules closely. Accordingly, replacements of an amino acid at the surface are more likely to destroy function than the same replacement in a freely soluble protein such as hemoglobin, or in a part of a protein that is cleaved and discarded (such as fibrinopeptide) at the time when the protein is stimulated to become an active enzyme. Proteins that tolerate amino acid substitutions more readily usually evolve faster. Evolutionary relationships among closely related organisms can be inferred by studying rapidly evolving proteins, whereas slowly evolving ones are used to reconstruct the evolutionary history of remotely related organisms.

Recent developments in molecular biology have made it easy to obtain the nucleotide sequences of genes and other DNA segments. Al-

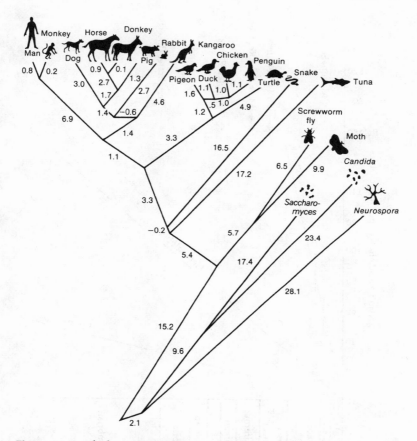

Figure 11.4 Phylogeny of twenty organisms, based on the cytochrome *c* data matrix (see Table 11.2). The numbers on the branches are the estimated number of substitutions that have taken place in the gene coding for the cytochromes *c*. (Fractions appear on the branches because they provide the best fit to the data and thus are the best estimates, although clearly substitutions occur in integer numbers.) This phylogeny agrees fairly well with phylogenetic relationships inferred from the fossil record and other sources. The agreement is remarkable, since it is based on the study of a single protein and encompasses organisms ranging from yeast, through insects, fish, reptiles, amphibians, birds, and mammals, to humans. (From F. J. Ayala and J. W. Valentine, *Evolving: The Theory and Processes of Organic Evolution* [Menlo Park, Calif.: Benjamin/Cummings, 1979]; after W. M. Fitch and E. Margoliash, "Construction of Phylogenetic Trees," *Science* 155 [1967]: 279–84.)

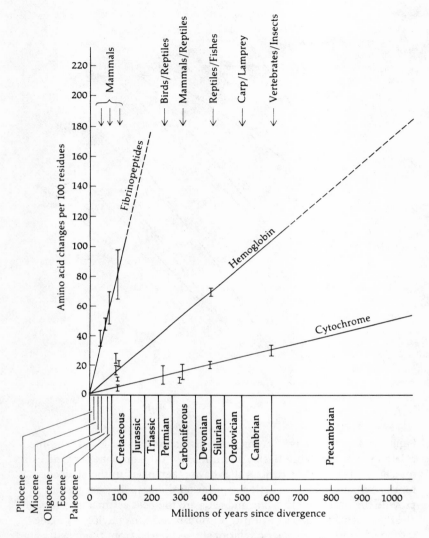

Figure 11.5 Rates of molecular evolution in three different proteins. The fibrinopeptides evolve rapidly, the cytochromes *c* slowly, and the hemoglobins at an intermediate rate. (From F. J. Ayala, *Population and Evolutionary Genetics: A Primer* (Menlo Park, Calif.: Benjamin/Cummings, 1982); after R. E. Dickerson, "The Structure of Cytochrome *c* and the Rates of Molecular Evolution," *J. Mol. Evol.* 1 [1971]: 26.)

though the number of genes that have been sequenced is still relatively small, new sequences are being obtained at a high rate, and a growing research program (the Human Genome Project) is promoting such work on a variety of organisms. The nucleotide sequence of genes provides more evolutionary information than the amino acid sequence of proteins, because the same amino acid may be coded for by several different codons (sequences of three nucleotides). Indeed, it is in these nucleotide sequences that the evolutionary and genetic information resides, whereas proteins are expressions of this information.

Phylogeny: Species Versus Genes

A conspicuous feature of evolution is the emergence of new functions and increased complexity. One starting point is gene duplication, the presence within the same organism of two or more identical or very similar copies of the same gene. Gene duplication represents a dimension of the evolutionary process that was completely opaque to the traditional methods of paleontology and comparative anatomy. Molecular biology has now revealed widespread gene duplication. Even more, it has revealed that there are within an organism numerous families of genes more or less similar in sequence to each other as a consequence of the time elapsed since duplicates diverged from the gene ancestral to the family.

An excellent example of a gene family is provided by the genes coding for hemoglobins and myoglobins in vertebrates. These proteins are involved in the transport and storage of oxygen—myoglobin in muscle, hemoglobin in blood. Their evolution can be traced to a single gene that became duplicated more than 650 million years ago in the ancestral line of the vertebrates. The duplication of that ancestral gene made possible the evolution of two types of proteins specialized for the particular demands of muscle and of blood.

Myoglobins consist of a single protein chain ("polypeptide") arranged in a complex three-dimensional structure with a heme group—a nonprotein molecule, with an iron atom, that plays a key role in the association and dissociation with oxygen and carbon dioxide essential for respiration. In the primitive fishes called cyclostomes (fishes without jaws, like lampreys and hagfishes), the hemoglobins also consist of only one polypeptide. Hemoglobin molecules in most vertebrates, however, consist of four subunits: two polypeptides of one kind (called alpha) and two of another (called beta), each with a heme group in its center. The gene coding for hemoglobin became duplicated some 500 million years ago in a fish lineage from which descend all modern fishes

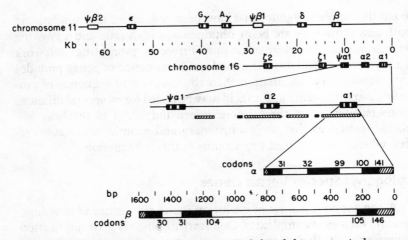

Figure 11.6 Chromosomal arrangement of the globin genes in humans. *Top.* The beta-like genes (β) are on chromosome 11; the alpha-like genes (α) on chromosome 16. *Middle.* The α1 and α2 genes are the result of a duplication and are nearly identical in nucleotide sequence. In addition, the nucleotide sequence that separates ψα1 from α2 is largely similar to the sequence between α2 and α1, as indicated by identical shading in the blow-up below the middle map. *Bottom.* The α and β genes. All other globin genes have the same organization: three exons (i.e., coding regions, *in black*) separated by two introns (intervening sequences, *in white*). The length of the introns is variable, but their position is the same—they are located between functionally identical codons. The conservation of the number and position of the introns throughout the hundreds of thousands of years of evolution of the globin genes indicates that the introns play a functional role, still not well understood. The scale on top is in kilobases (Kb) or thousands of base pairs; the scale at bottom is in number of base pairs (bp).

(except the cyclostomes) and all other vertebrates. The products of that gene duplication, the alpha and the beta globin genes, made possible the evolution of the tetramer hemoglobin molecules (made up of two alpha polypeptides and two beta polypeptides) found in most vertebrates. Tetramer hemoglobins are much more efficient for oxygen transport than the single-polypeptide hemoglobins of the cyclostomes.

The story of the globin genes does not end there. The alpha and the beta genes each became duplicated again, more than once at different times in vertebrate history. Modern hemoglobins, as they exist in humans, for example, always consist of two alpha-like identical subunits and two beta-like identical subunits. In human adults the beta-like subunits are of two kinds (beta and delta), so that adults have two kinds of hemoglobin. Moreover, humans are endowed with yet more hemoglobin modulations, such as hemoglobins adapted to the special needs of the first eight weeks of embryonic life or to those of the later fetus (see Table 11.3, Figures 11.6, 11.7).

Nothing of this fascinating story of successive gene duplications and functional specialization was known, or could have been known, before molecular biology came about. The evolution of life can now be

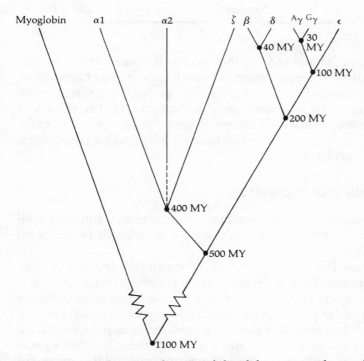

Figure 11.7   Evolutionary history of the globin genes. The symbols on top of each lineage represent the same genes as those in Figure 11.6. The dots indicate where the ancestral genes were duplicated, giving rise to a new gene line. The approximate times when the duplications occurred are indicated in millions of years (MY). The time when the duplication of the α1 and α2 genes occurred is unknown, but is not earlier than 400 MY. All the hemoglobin genes derive from a duplication that occurred about 1100 MY, from which the myoglobin gene line also derives. (From F. J. Ayala and John A. Kiger, Jr., *Modern Genetics* [Menlo Park, Calif.: Benjamin/Cummings, 1984]).

Table 11.3   Hemoglobins present at different stages of human life, their tetramer structure, and the genes encoding the corresponding polypeptides

	Hemoglobin	Tetramer Structure	Genes
Adult	A	$\alpha_2\beta_2$	α1 and α2, β
	A_2	$\alpha_2\delta_2$	α1 and α2, δ
Fetus	F	$\alpha_2\gamma_2$	α1 and α2, $^G\gamma$ and $^A\gamma$
Embryo (to 8 weeks)	Gower 1	$\zeta_2\epsilon$	ζ1 and ζ2, ϵ
Transition from embryo to fetus	Gower 2	$\alpha_2\epsilon2$	α1 and α 2, ϵ
	Portland	$\zeta_2\gamma_2$	ζ1 and ζ2, $^G\gamma$ and $^A\gamma$

studied with a richness of detail that was unsuspected two decades ago. It is not only that so much more can now be known, and known more precisely, about the historical relationships among organisms, but also that fundamentally new discoveries are being made. The process of gene duplication is one such discovery. Similarly, how primitive genes came about in the first place during the early history of life is just beginning to be elucidated.

Some Additional Implications

Molecular genetic studies are rapidly enriching evolutionary biology in many additional ways, only some of which will be briefly reviewed here.

The stark Darwinian picture of our origin still encounters enormous skepticism. Though the evidence for the evolutionary continuity of all organisms is widely accepted, the idea that it occurred through a natural, unguided process requiring billions of years is another matter. Virtually all biologists recognize this mechanism as fundamental for the unifying, central role of evolution in biology today, and they see it as enormously strengthened by the evidence from molecular genetics. But in a poll of a random sample of the population in this country, only 9 percent of respondents believed in an unguided process.[3]

The main reason surely is conflict with religious beliefs—a fundamentalist interpretation of the Bible for many, but more broadly also the conviction that evolution undermines assumptions required for psychic well-being and for justifying a moral consensus. But there is another, very different reason for skepticism: common sense conflicts with the claim that such complex organs as the eye could have evolved through a series of tiny, random, undirected steps.

Molecular genetics has now greatly advanced our understanding of what these "tiny steps" may be. It has revealed an extraordinary range of changes included in the concept of mutation—a term used here in the broad sense of any one-step change in DNA sequence. The most obvious kind of mutation, the replacement of one nucleotide by another through an error in DNA replication, results in some cases but not others in the replacement of an amino acid. If it does, it may eliminate or reduce the activity of an enzyme, or it may alter the range of compounds on which the enzyme acts. If the mutation does not result in a change in an amino acid, it may be "neutral" (i.e., have no effect on structure or function) but nevertheless contribute to evolution by broadening the base for subsequent mutations at the same site.

We now recognize many kinds of single-nucleotide changes with much larger effects than those just described. While mutation in a gene for an enzyme directly affects only that product, a mutation in a regulatory sequence of DNA may have a broader effect, given that a common regulatory mechanism often governs the formation of many proteins. Similarly, when a change in a single nucleotide alters the site ("punctuation") where a gene begins or ends its transcription, it may add or remove a substantial string of amino acids. Finally, when a mutation adds or removes one nucleotide within a gene, it shifts the normal frame in which the ensuing DNA is read as a succession of trinucleotide codons. This frameshift changes all the subsequent codons in the gene, and hence all the corresponding amino acids in the resulting protein.

In addition to these minimal changes in DNA sequence, which may have either small or large effects on the phenotype, evolution also builds on much larger changes in the DNA. Large blocks of DNA can be transferred between organisms, or within an organism to a new location on a chromosome where the regulation will be different. The duplication of genes, discussed above, is a special case of gene translocation. Moreover, viruses, and the closely related autonomous units of DNA called plasmids, can transfer blocks of DNA between organisms, and the foreign DNA can be incorporated into a host chromosome.

In most genes of higher organisms only part of the DNA sequence, the exons, is translated into the sequence of amino acids making up the polypeptide. Intervening sequences, the introns, lie between the exons and are spliced out of the initial RNA transcript before amino acid translation occurs. The different exons within a gene often code for polypeptide sequences that each serve as a unit with a specific function in the protein (e.g., affecting its specificity, location, or regulation). The introns serve as major sites of genetic recombination between two homologous chromosomes, and also as sites of excision for transfer of a DNA block. They thus facilitate duplication of whole genes or simply of functional parts. This process accelerates the formation of potentially useful novel proteins by combining successful pieces of earlier ones.

The processes just surveyed illustrate that evolution can tap a great variety of mechanisms for generating novelty, differing widely in the magnitude of their effect on the genotype. Moreover, small effects on the genotype may result in rather large effects on the phenotype—and the phenotype is what natural selection acts on. These findings, together with developments in other areas of evolutionary biology, make it more plausible that the observed small changes of "microevolution" (changes within species) can be extrapolated to account for the larger changes in "macroevolution" (the changes needed for the emergence of new species and new organs).

Evolutionary biology is also influenced by our growing understanding of how a particular sequence of amino acids in a polypeptide chain folds optimally into the specific three-dimensional structure of a functional protein. This folding provides the specific sites on the surface of the protein that interact with other molecules. The study of molecular evolution has, consequently, become concerned not only with the homology between protein sequences, but also with relationships of shape within families of proteins with related functions.

The sequence-dependent specific adhesions of proteins with other molecules, long known in the actions of enzymes and antibodies, also lead to the aggregations by which macromolecules form cells, tissues, organs, and the whole organism. This field, molecular morphogenesis, has become one of two major areas of study in developmental biology. The other is the study of the regulatory mechanisms that turn different genes on or off in specific cells at different stages of development. Since the processes of development, from the fertilized egg to the final form of a higher organism, place major constraints on the range of changes in the organism that can survive in evolution, the emerging field of molecular developmental biology will become increasingly significant for evolutionary biology.

The study of evolution, like the study of genetics, is rapidly moving from formal relations to concrete mechanisms. For example, in studying the evolution of the physical and mental differences between chimpanzee and human we are now concerned not only with the time scale revealed by the fossil record. The challenge for the future is to understand the differences in specific genes, adding up to only 1 percent overall difference in DNA sequence, that account for the gap between humans and the higher apes.

Finally, we might be reminded here of a point that was implicit in some of the earlier chapters in this book: the controversy over hypothetical dangers from engineered organisms, especially bacteria, has given evolutionary biology a new importance for practical and political affairs. For this is the field that encompasses study of the survival, spread, or extinction of organisms in nature. Moreover, the scientific discipline most directly concerned with the spread of infectious disease, epidemiology, is a kind of applied evolutionary biology. This consideration provides additional reasons to hope that evolutionary biology will receive the attention in our schools that it deserves, despite the efforts of some religious fundamentalists to block this teaching.

Notes

1.   The task of reconstructing evolutionary history from the study of living organisms is far from simple. For example, resemblances due to common descent must be set apart form resemblances due to similar ways of life, to life in the same or similar habitats, or to accidental convergence of forms. The reconstruction of evolutionary history from fossil remains encounters similar difficulties, plus others arising from incomplete preservation, problems of dating, and so on.

2.   W. M. Fitch and E. Margoliash, "Construction of Phylogenetic Trees," *Sicence* 155 (1967): 279–84.

3.   *New York Times*, 29 August 1982, p. 22.

12

Regulation

Henry I. Miller

To date, for the regulation of the new biotechnology this has been the best of times and the worst of times. On the positive side, oversight by the National Institutes of Health has been enlightened, with progressively decreasing stringency of their guidelines for laboratory research. The approach of the Food and Drug Administration (FDA) to the new products, treating them in the same way as similar products made in other ways, has been validated for about a thousand products, with more than 350 approved for marketing and approximately 750 others in clinical trials. On the negative side, planned introductions ("deliberate releases") have frequently been overregulated. Several European countries have been unnecessarily restrictive, and they have paid the price in diminished research and development activity. Even in the United States, an intermittently inhospitable regulatory climate and a lack of public understanding have discouraged research and development in certain areas. In particular, academic and industrial research has moved away from field trials of recombinant organisms, and investor interest in this sector has diminished.

The emergence and persistence of hostile regulatory climates can be traced to the propagation of various myths about the new biotechnology, not balanced by vigorous efforts at public education. In fact the vast majority of applications of the "new biotechnology" are fundamentally similar, on both theoretical and experiential grounds, to well-accepted, ubiquitous applications of the "old biotechnology," and they are demonstrably safe. Treating them as hazardous will ultimately cause the public to suffer from a failure to develop new products.

The goal of the U.S. government's coordinated approach to the regulation of biotechnology is to limit potential risks while encouraging the innovation, development, and availability of new products.[1] However, not only must products be safe, but the public must have confidence in their safety. In this respect, the advent of the new biotechnology poses a major challenge that involves perception as much as reality. Only if we can correct the misconceptions and eliminate the harmful myths that have grown up around the buzzword *biotechnology* can its potential be realized.

Myth Number One: Biotechnology Is a Discrete Entity

One myth is that biotechnology is something discrete or homogeneous, a corollary of which is that there exists *a* "biotechnology industry" that can or should be rigidly controlled. This view is facile but inaccurate. *Biotechnology* is merely a catchall term for a broad group of useful, enabling technologies with wide and diverse applications in industry and commerce. A useful working definition used by several government agencies is the application of biological systems and organisms to technical and industrial processes. This definition encompasses processes as different as fish farming, forestry, the production of enzymes for laundry detergents, and the genetic engineering of bacteria to clean up oil spills, kill insect larvae, or produce insulin.

Biotechnology is thus myriad dissimilar processes, producing even greater numbers of dissimilar products for vastly dissimilar applications. These processes and products have so little in common with one another that it is difficult to construct valid generalizations about them, for whatever purpose. Because of this lack of systematic, uniform characteristics, effective legislation or oversight cannot be as homogeneous as is possible, for example, for the underground coal-mining industry.

The diversity of biotechnology dictates that the regulation of so many end uses must be accomplished by many government agencies, differing in their jurisdiction and the nature of their evaluation of the products. Clearly, for example, a review by the Environmental Protection Agency (EPA) of an enzyme used as a drain cleaner will be different from a review by the FDA of the same enzyme injected into patients to dissolve blood clots.[2] The diversity of products and their applications argues against the usefulness of legislation or regulations that attempt to encompass unnatural groupings such as "biotechnology," "genetically manipulated organisms," or "planned introductions."[3]

Myth Number Two: Biotechnology and Genetic Engineering Are New

A second myth is that biotechnology is new. On the contrary, many forms of biotechnology have been widely used for millennia. Earlier than 6000 B.C. the Sumerians and the Babylonians exploited the ability of yeast to make alcohol by brewing beer. A "picture" of the ancients preparing and fermenting grain and storing the brew has actually been retained for posterity in a hieroglyphic.[4] An even older subset of biotechnology is domestication, the indirect manipulation of an organism's genes by the guided mating of animals and crop plants so as to enhance desired characteristics. In this process changes are achieved at the level of the whole organism, by selection for desired phenotypes; the genetic changes, often poorly characterized, occur concomitantly. The new technologies, in contrast, enable genetic material to be modified at the cellular and molecular levels. These new techniques yield more precise and deliberate variants, and they consequently produce better-characterized and more predictable results. Nevertheless, the aims of molecular and of classical domestication are the same.

Building on the domestication of microorganisms for the modification of foods, industry has used classical genetic methods to yield many valuable organisms. One example is the genetic improvement of *Penicillium chrysogenum*, the mold that produces penicillin. During the past several decades the yields have been increased more than a hundredfold by several methods, including the screening of thousands of isolates and the use of mutagens to accelerate variation. There are many similar examples; microbial fermentation is employed throughout the world to produce a variety of important substances, including industrial detergents, antibiotics, organic solvents, vitamins, amino acids, polysaccharides, steroids, and vaccines.[5] The value of these and similar products of conventional biotechnology exceeds $100 billion annually. Today, with the identification of the genes responsible for the biosynthesis of various antibiotics, the empirical search for improved products or yields is being supplanted by the introduction of specific, directed genetic changes.

In addition to such contained industrial applications, there have been many beneficial planned introductions of other organisms into the environment, including insects, bacteria, and viruses. Insect release was used successfully to control troublesome weeds in Hawaii in the early twentieth century. Other examples are the successful program for the biological control of St.-John's-wort (Klamath weed) in California by insects in the 1940s and 1950s, and the more recent use of an introduced rust pathogen to control rush skeletonweed in Australia. This area of research, the biological control of weeds, is actively promoted by the U.S. Departments of Interior and Agriculture.[6]

Currently, more than a dozen microbial pesticidal agents are approved and registered with the EPA, and these organisms are marketed in hundreds of different products for use in agriculture, forestry, and insect control.[7] In another major area, preparations containing the bacterium Rhizobium, which enhances the growth of leguminous plants such as soybeans, alfalfa, and beans, have been sold in this country since the late nineteenth century. These products allow the plants to produce nitrogen fertilizer from the air.

The most ubiquitous planned introductions of genetically modified organisms have involved the vaccination of human and animal populations with live, attenuated viruses. Live viruses modified by various techniques and licensed in the United States include mumps, measles, rubella, poliovirus, and yellow fever. Inoculation of a live viral vaccine entails not only infection of the recipient but sometimes the possibility of further transmission of the virus in the community with the presumed risk of serial propagation. Nevertheless, none of the vaccine viruses has become established in the environment, despite being present there. For example, vaccine strains of poliovirus are present in sewage in the United States and the United Kingdom, but this reflects the continuing administration and excretion of live virus vaccine rather than its serial propagation in the community.[8]

Viral vaccines produced with older genetic techniques for eliminating their virulence have been awesomely effective throughout the world and have completely eradicated the dread smallpox virus. Despite rare adverse reactions, the vaccines are rivaled, as a promoter of human longevity and quality of life, only by the agricultural "green revolution." The newest biotechnological techniques, including recombinant DNA, are already providing still more precise, better-understood, and more predictable methods for manipulating the genetic material of viruses and higher microorganisms for this purpose.

Myth Number Three: The Unknowns Outweigh the Knowns

It is excessively negative to argue that "as far as the ecological properties of microbes are concerned, the unknowns far outweigh the knowns."[9] One could just as easily substitute "the functions of mutations in the poliovirus vaccine"; but for four decades these uncharacterized genetic changes have not prevented our using the empirically developed, live, attenuated poliovirus vaccine. The statement is particularly dubious for many microorganisms of commercial interest, including *Pseudomonas syringae*, Thiobacillus species, *Bacillus thuringiensis*, *Bacillus subtilis*, Rhizobium, and baculovirus, as well as the microbes used in fermentation to produce bread, cheese, wine, beer, and yogurt. In fact, the vast majority of microbes are essential to ecosystem processes or otherwise beneficial to mankind, and only a minuscule fraction are pathogenic or otherwise harmful.

Until the recent furor over engineered organisms, small-scale field trials of microbes modified by classical techniques were long exempt in the United States from both the pesticide and toxic substances regulations, and the safety record is admirable. The scientific and prior experience, when applied logically to risk assessment, do enable us to make useful predictions.

Myth Number Four: Novel and Dangerous Organisms Will Be Created

The degree of novelty of microorganisms or macroorganisms created by the new genetic engineering techniques has been wildly exaggerated. A corn plant that has newly incorporated the gene for and synthesizes the *Bacillus thuringiensis* toxin is still, after all, a corn plant. *Escherichia coli* that has been programed to synthesize human alpha-interferon by means of recombinant DNA techniques differs very little from its unmanipulated siblings that manufacture only bacterial molecules. Moreover, nature has already tried out innumerable recombinations between even very distantly related organisms via several mechanisms.[10] Bacteria can take up naked DNA, though inefficiently, and in nature they have long been exposed to DNA from disintegrating mammalian cells, as in the gut, in decomposing corpses, and in infected wounds. The human population alone excretes on the order of 10^{22} bacteria per day; hence, over the past million years innumerable mammalian bacterial hybrids are likely to have appeared and been tested and discarded by natural selection. An analogous argument can be made for recombination among fungi, bacteria, viruses, and plants.

Both genetic and ecological constraints operate to prevent the emergence of exceedingly pathogenic microbial variants, though even a single-point mutation (i.e., in one nucleotide) can alter virulence.[11] And while the variations that continually arise on an enormous scale in nature do occasionally produce a modified pathogen, such as an influenza virus with increased virulence or the AIDS virus, the chances of such an event arising from the small-scale changes made by human intervention do not compare with the tremendous "background noise" in nature.

Since evolution continually creates novelty, the distinction between *natural* and *unnatural* (or *novel*) is not a clear one and is, arguably, irrelevant. *Novel* is not synonymous with *dangerous*. Nor need it even imply uncertainty if it involves familiar components.

Myth Number Five: Nonpathogens Will Be Transformed into Pathogens

A frequent concern is that genetic manipulation may inadvertently transform a nonpathogen into a pathogen. However, this view ignores the complexity and the multifactorial nature of pathogenicity. Pathogenicity is not a trait produced by some single omnipotent gene; rather, its evolution in a nonpathogen requires emergence of a special set of properties that involve a number of genes.

A pathogen must possess two general characteristics, which are themselves multifactorial. First, it must be able to metabolize and multiply in or upon host tissues; that is, the oxygen tension and pH must be satisfactory, the temperature suitable, and the nutritional milieu favorable. Second, assuming an acceptable range for all of the many conditions necessary for metabolism and multiplication, the pathogen must be able to resist the host's defense mechanisms for a period of time sufficient to reach the numbers required actually to produce disease.

The organism must be meticulously adapted to its pathogenic lifestyle. Even a gene specifying a potent toxin will not convert a harmless bacterium into an effective source of epidemic, or even localized disease, unless many other required traits are present. These include, at the least, resistance to enzymes, antibodies, and phagocytic cells in the host; the ability to adhere to specific surfaces; and the ability to thrive on available nutrients provided by the host. No one of these traits confers pathogenicity, though a mutation that affects any essential one can eliminate it.

Another important consideration is that severe pathogenicity and epidemicity are more demanding of favorable conditions, and much rarer in nature, than mild degrees of pathogenicity. Hence, the probability of inadvertently creating an organism capable of producing a medical catastrophe must be vanishingly small.

Myth Number Six: All Technology Is Intrinsically Dangerous

Another myth is that the application of all new technology is dangerous. The basis of this view may be the atavistic fear of disturbing the natural order and of breaking primitive taboos, combined with unfamiliarity with the statistical aspects of risk. Promulgators of this myth cite the hazards of toxic chemical waste dumps and the technical problems of the nuclear industry, but they conveniently ignore the overwhelming successes of telephonic communication, vaccination, blood transfusion, microchip circuitry, and the domestication of animals, plants, and microbes. It is worth remembering that critics predicted electrocution from the first telephones, the creation of human monsters by Jenner's early attempts at smallpox vaccination, and the impossibility of matching blood for transfusions. They said, in effect, "The costs will be too high; there is no such thing as a free lunch."

No responsible person would deny that novel practices, processes, or products of biotechnology could be hazardous in some way. Some of these hazards are already well known: workers purifying antibiotics have experienced allergic reactions; beekeepers have been stung; laboratory workers have inadvertently been infected by pathogenic bacteria; and vaccines have occasionally elicited serious adverse reactions.

In addition, introductions of exotic species, such as English sparrows and gypsy moths, have had serious economic consequences. An analogy is sometimes made between the introduction of these non-native (alien or "exotic") organisms and the possible consequences of field trials with organisms modified by the new techniques. However, the comparison is largely specious, relying on the assumption that recombinant DNA manipulations can alter the properties of an organism in a wholly unpredictable way that will cause it to affect the environment adversely. Both theory and experience make this outcome very unlikely.

Generally, introductions will be of indigenous organisms that differ minimally, often by only the insertion or deletion of a single structural gene, from organisms already present in the environment, and that will not enjoy a selective advantage over their wild-type cohorts. They will be subject to the same physical and biological limitations of their environments as their unmodified parents. Thus, as a model for organisms modified by the techniques of the new biotechnology, the older system of selective breeding and testing of domesticated plants, animals, and microbes, with its long history and admirable safety record, is more applicable than the introduction of exotic organisms.

In any case, a complex and comprehensive regulatory apparatus based in numerous federal agencies in the United States has long provided reasonable supervision of the safety of food plants and animals, pharmaceuticals, pesticides, and other products that can also be produced by biotechnology. There is every reason to expect that it will continue to be equal to the task and to perform in a way that does not stifle innovation. There may never be a "free lunch," but often we can make it an excellent value.

Risk Assessment for the New Biotechnology

Among those who are scientifically knowledgeable about the new methods of genetic manipulation there is wide consensus, based on the large body of knowledge that has been gathered about the biological world, that existing risk assessment methods are suitable for environmental applications of the new products. Several appropriate procedures are available; and while it is true that risk assessment is not an exact, predictive discipline, the National Science Foundation has concluded that available methods provide both a useful foundation and "a systematic means of organizing a variety of relevant knowledge about the behavior of microorganisms in the environment."[12] In assessing risks of the new biotechnology, the need is not for new methods but rather for ways of ascertaining the correct underlying assumptions. For risk assessment, as for many other aspects of the new biotechnology, the new products manufactured with the new processes do not necessarily require new regulatory or scientific paradigms.

The U.S. National Academy of Sciences has developed a landmark policy statement on the planned introduction of organisms into the environment, which has had wide-ranging impacts in the United States and internationally. Several of its most significant conclusions and recommendations are as follows:

1. Recombinant DNA techniques constitute a powerful and safe new means for the modification of organisms.
2. Genetically modified organisms will contribute substantially to improved health care, agricultural efficiency, and the amelioration of many pressing environmental problems that have resulted from the extensive reliance on chemicals in both agriculture and industry.
3. There is no evidence of the existence of unique hazards either in the use of recombinant DNA techniques or in the movement of genes between unrelated organisms.
4. The risks associated with the introduction of recombinant DNA–engineered organisms are the same in kind as those associated with the introduction of unmodified organisms and organisms modified by other methods.
5. The assessment of risks associated with introducing recombinant DNA organisms into the environment should be based on the nature of the organism and of the environment into which it is to be introduced, and independent of the method of engineering per se.[13]

In an expansion of this policy, the National Research Council (NRC) has stated that "no conceptual distinction exists between genetic modification of plants and microorganisms by classical methods or by molecular techniques that modify DNA and transfer genes," whether in the laboratory, in the field, or in large-scale environmental introductions. This formulation has three implications. First, the *product* of genetic modification and selection constitutes the primary basis for decisions about the environmental introduction of a plant or microorganism, and not the *process* by which the product was obtained. Second, although knowledge about the process used to produce a genetically modified organism is important in understanding the characteristics of the product, the nature of the process is not a useful criterion for determining how much oversight the product requires. Third, organisms modified by modern molecular and cellular methods are governed by the same physical and biological laws as organisms produced by classical methods, with which there is a vast body of experience. The NRC proposed that the evaluation of experimental field testing be based on three considerations: familiarity, that is, similarity to introductions with a safe history; ability to confine or control the spread of the organism; and likelihood of harmful effects if the organism should escape control or confinement.[14]

The correct paradigm for regulation of products of the new genetic engineering may be summarized in a syllogism. Industry, government, and the public already have considerable experience with the planned introduction of traditional genetically modified products for both agriculture and live vaccines. Existing regulatory schemes have effectively protected human health and the environment without stifling industrial innovation. There is no evidence that unique hazards exist either in the use of recombinant DNA techniques or in the movement of genes between unrelated organisms. Therefore, there is no need for additional regulatory mechanisms to be superimposed on recombinant DNA regulation.

The syllogism assumes, of course, that prior to the advent of the new genetic engineering, the government exercised adequate control over the testing and use of living organisms and their products. This assumption is certainly questionable. Nevertheless, in the United States scientists have had a high degree of unregulated freedom of experimentation, with pathogens as well as nonpathogens, indoors and out; the resulting harmful incidents have been few; and the benefits, both intellectual and commercial, have been substantial.

Implications for Government Policy

Of concern to many practitioners and regulators of biotechnology are two nonscientific issues that could play a dominant role in the future of biotechnology development and use in the United States and abroad. First, the regulatory climate could, if irrational or inappropriately risk-averse, impede industry's eventual introduction of products now being developed. This obstacle could lead to future withdrawal or diminution of new product development, with continued reliance on less sophisticated, and often more hazardous, alternative technologies. Second, concerns within the financial community about the long-term stability and success of companies doing business in environmentally regulated fields could depress the values of those companies, drying up capital for the continued development and testing necessary to satisfy regulatory requirements.[15]

The requirements of U.S. regulatory agencies remain, after years, in various stages of readiness. The last major coordinated statement of policy by the federal agencies was made in 1986.[16] Since then, the EPA has been attempting to refine and clarify that preliminary proposal.[17] At the Department of Agriculture, policies at the Animal and Plant Health Inspection Service have remained essentially unchanged since 1987, while the research side of the department is currently developing guidelines for research.[18] The FDA has imposed no new procedures or requirements for products made with the new biotechnology.[19]

Those who are associated with the regulatory issues of biotechnology have a critical responsibility to act quickly and definitively to balance the various opposing forces. The notion that the products derived from the new biotechnology defy useful, accurate risk assessment or are too dangerous to introduce into the environment is without merit: both theory and experience repudiate this assertion. Policy must be guided in part by an understanding that there are genuine costs of overly risk-averse regulatory policies that *prevent* the testing and approval of new products. For example, the destruction of crops by frost while "Ice-minus" bacteria languish untested represents a significant toll, as does the continued application of chemical pesticides that could be replaced by new biorational pesticides. Bengt Jonsson is correct in stating that safety can be bought only at a cost.[20]

Figure 12.1 illustrates the relationship between safety and resources of manpower, capital, and raw materials expended. At a given stage of technology an increase in the safety level from S_1 to S_2 can be achieved only through an increase in resources from R_1 to R_2. When the FDA, for example, requires additional clinical or animal toxicity data from a manufacturer seeking marketing approval for a product, the agency has

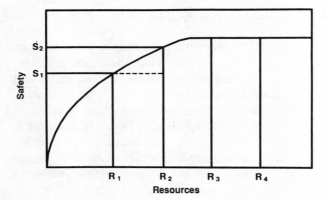

Figure 12.1 Relationship between safety produced and resources expended. The expenditure of resources such as money and manpower is shown on the abscissa, the resultant safety on the ordinate. (After B. Jonsson, "The Economic Impact," in IFPMA symposium, *Scientific Innovation in Drug Development: The Impact on Registration* [Geneva: IFPMA, 1986].)

made the judgment that its evaluation of the product is on this portion of the curve. On the flat portion of the curve, for example moving from R_3 to R_4, no additional safety advantage is obtained by increasing the expenditure of resources above R_3. The outlay of resources by both industry and government is ultimately borne by the consumer, and overregulation that discourages innovation and reduces competition delays or denies the benefits of the new products to the public.

The discussion of overregulation is more than theoretical discourse. Internationally, for example, regulation has been a hodgepodge. Although Japan has adopted an illogical process-based regulatory approach, the resulting restrictions have been annoying rather than debilitating, and the new biotechnology applied to pharmaceuticals there has sustained significant progress. By contrast, no Japanese field trials of living organisms manipulated with the new biotechnology have been carried out, and research and development in this area appear minimal.

Denmark and West Germany have been the prototypical overregulators, unyielding gatekeepers who have kept virtually any new biotechnology products from emerging. Denmark's Novo Industri, the world's largest producer of enzymes, has moved its research and development to Japan, and Germany has experienced a veritable hemorrhage of research and development.[21] Bayer AG has begun to establish research and development facilities abroad, especially in the United States. BASF has shelved investment plans for a $70-million research and development facility in Ludwigshafen and will erect it in Boston instead. Hoechst has decided to locate its international center for biotechnology products in Japan. Some German companies have given up the ghost completely. A midsized company, Wiemer Pharmazeutisch-Chemische Fabrik, has abandoned its biotechnology production and liquidated its equipment; Abbott Diagnostics in Wiesbaden has postponed indefinitely its biotechnology plans.

The European Economic Community (EEC) has begun the implementation of extremely regressive directives for both contained uses and planned introductions of genetically modified organisms into the environment.[22] Strongly process-based, the directives' underlying premises are scientifically flawed; they would hinder research and development activities throughout the EEC; and their provisions could potentially be used to erect nontariff trade barriers to foreign products.[23]

Investor interest in two sectors of U.S. industry that use new biotechnology, namely biopharmaceuticals and agriculture, provides another measure of the effect of overregulation (see Figure 12.2). Over the period 1983–88 investor interest in the biopharmaceutical sector was strong, performing about equal to the market. By contrast, the agricultural sector performed dismally and lagged far behind biopharmaceuticals. One reason for this disparity is certainly the marked difference in the regulatory milieu. While the FDA has approved more than a dozen therapeutics and vaccines in less than half the usual approval times and has permitted more than eight hundred clinical trials of such products, field testing of new biorational pesticides and other agricultural products that fall under the EPA's oversight have faced much-publicized regulatory and judicial delays. Private sector investors appear unwilling to gamble funds on research and development in an area where the climate is inhospitable—where it can take years as well as millions of dollars simply to perform a single field test, let alone meet the requirements for marketing approval down the road.

Another homegrown example of the problems of overregulation is to be found in Wisconsin, where by 1988 the prospect of months of regulatory paperwork to obtain approval for a single field trial was causing investigators, especially young, untenured academic researchers, to shy away from research areas that require field testing.[24] A 1989 survey of agricultural research in the public and private sectors uncovered a similar problem. Among the 371 respondents who work with genetically engineered organisms, approximately 10 percent of those in the public sector (including universities and government laboratories) and 20 percent

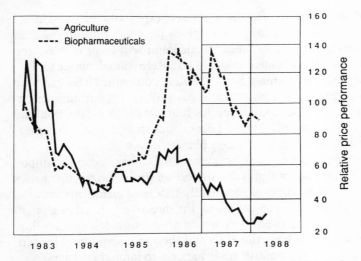

Figure 12.2   Relative performance of two sectors of U.S. industry using the new biotechnology. The share-weighted stock performance of companies normalized to Standard and Poor's 400 Index is grouped by biopharmaceuticals (forty companies) and agricultural (ten companies). The indices were set at 100 for January 1983. (Data from L. I. Miller, Paine Webber, Inc.)

in the private sector had elected not to proceed with field trials, although they had organisms ready to test; three-quarters of these researchers cited governmental regulation as the reason.[25]

Thus, whole areas of research that could provide useful products, ranging from growth promotants to pesticides to biological treatments for toxic wastes, are being systematically avoided. At least two companies in this business felt compelled in 1989 to eschew the newest and most precise techniques of genetic engineering in favor of scientifically inferior but bureaucratically favored—in other words, less stringently regulated—techniques.[26] These are the fruits of *dis*incentives to the application of the new biotechnology.

Additional, second-order effects of overly risk-averse messages extend more broadly to science and technology. A former FDA commissioner, Donald Kennedy, warned that vital research is already curtailed in research institutions in parts of the United States. He pointed out that the antiresearch movement, whether from concern for laboratory animals or from fear of scientific experimentation, had prevented the opening of Stanford's completed $13-million biomedical research facility and had declared war on others planned at the University of California and elsewhere. Kennedy asked where this opposition expects society, without such facilities, to find cures for AIDS and other scourges.[27]

The professional practitioners, regulators, and observers of the new biotechnology must strive to demystify it and to provide the proper perspective for the public, because it is the public that will benefit most. The stakes are high in terms of both economics and social benefit. In recent years the FDA has approved several products of new biotechnology that are medical milestones. These include alpha-interferons for the treatment of a lethal leukemia, a monoclonal antibody preparation for preventing the rejection of kidney transplants, a new-generation vaccine for the prevention of hepatitis, a clot dissolver for treating heart attacks, and a hormone that reduces the blood transfusion requirements for kidney dialysis patients. Among myriad other applications, biotechnology promises vaccines against scourges such as malaria, schistosomiasis, and AIDS, as well as new therapies that could ameliorate or cure for the first time such genetic diseases as sickle cell anemia and certain inherited immune deficiencies. Used for generation of new medicines, food plants, and animals, biotechnology could provide partial solutions to that trinity of despair—hunger, disease, and the growing mismatch between material resources and population.

Finally, far from constituting an imminent threat to the environment, application of the new biotechnology may well provide solutions to existing environmental problems. For example, more efficient crop plants require a smaller commitment of land to agriculture, freeing it for recreation or for a return to wilderness; plants can be engineered that diminish the overall need for chemical pesticides and fertilizers; microbes can be engineered to degrade toxic compounds; and biological control agents are likely to be more predictable and safer than chemicals.

Conclusion

Biotechnology has been widely applied for millennia, resulting in innumerable successful and beneficial uses, including introductions into the environment. The precision and power of the genetic manipulation of both macroorganisms and microorganisms have greatly increased in recent decades with the development of molecular genetics. The techniques of the new biotechnology are best viewed as extensions, or refinements, of older techniques for genetic manipulation and domestication. For these reasons, among those knowledgeable about the new methods there exists wide consensus that current risk assessment methods are suitable for products of the new biotechnology, including their environmental applications; the new processes do not necessarily require new regulatory paradigms. There is no sound scientific rationale for fear of the newest techniques of genetic engineering or, indeed, of nonpathogenic, nontoxigenic organisms that have been modified by any technique.

Moreover, there are genuine costs of adopting overly risk-averse regulatory policies, which prevent the testing and approval of new products. Such policies are anti-innovative and anticompetitive, and they delay the benefits of new products to the public. These policies both emanate from and feed into a pernicious antiscience movement that threatens basic scientific research as well. The long-term remedy must be improved public education about science and technology.

Max Planck said more than half a century ago that an important scientific innovation rarely makes its way by gradually winning over and converting its opponents; rather, its opponents gradually die out and the succeeding generation is familiarized with the idea from the beginning. It would be tragic if this were the fate of the new biotechnology, for the result would be prolonged delays in the widespread testing and use of new products of value for health, food production, and the economy.

Notes

1. *Federal Register* 51 (1986): 23302–93.
2. *Federal Register* 49 (1984): 50856.
3. H. I. Miller, "FDA and the Growth of Biotechnology," *Pharma. Eng.* 6 (1986): 28.
4. A. L. Demain and N. A. Solomon, "Industrial Microbiology," *Sci. Am.* 245 (1981): 66–75.
5. "Health Impact of Biotechnology: Report of a WHO Working Group," *Swiss Biotech.* 2 (1985): 7–16.
6. D. L. Klingman and J. R. Coulson, "Guidelines for Introducing Foreign Organisms into the U.S.A. for Biological Control of Weeds," *Plant Dis.* 66, 12 (1982): 1205–9.
7. F. Betz, M. Levin, and M. Rogul, "Safety Aspects of Genetically Engineered Microbial Pesticides," *Recomb. DNA Tech. Bull.* 6 (1983): 135–40.
8. E. D. Kilbourne, "Epidemiology of Viruses Genetically Altered by Man—Predictive Principles," in Banbury Report 22, *Genetically Altered Viruses and the Environment* (Cold Spring Harbor Laboratory, N.Y., 1985).
9. F. E. Sharples, "Regulation of Products from Biotechnology," *Science* 235 (13 March 1987): 1329.
10. B. D. Davis, "Evolution, Epidemiology, and Recombinant DNA," *Science* 193 (1976): 442.
11. Kilbourne, "Epidemiology of Viruses."
12. V. T. Covello and J. R. Fiksel, eds., *The Suitability and Applicability of Risk Assessment Methods for Environmental Applications of Biotechnology* (Washington, D.C.: National Science Foundation, 1985).

13. National Academy of Sciences, *Introduction of Recombinant DNA–Engineered Organisms into the Environment: Key Issues* (Washington, D.C.: National Academy Press, 1987).

14. National Research Council, *Field Testing Genetically Modified Organisms: Framework for Decisions* (Washington, D.C.: National Academy Press, 1989).

15. D. T. Kingsbury, *Bio/Technology* 4 (1986): 1071-74.

16. *Federal Register* 51 (1986): 23302–93.

17. *Federal Register* 54 (1989): 7027.

18. *An Analysis of the Federal Framework for Regulating Planned Introductions of Engineered Organisms*, Interim Summary Report, Keystone National Biotechnology Forum, The Keystone Center, Keystone, Colorado, February 1989.

19. H. I. Miller and F. E. Young, "FDA and Biotechnology: Update 1989," *Bio/Technology* 6 (1988): 1385–92.

20. B. Jonsson, "The Economic Viewpoint," in IFPMA symposium, *Scientific Innovation in Drug Development: The Impact on Registration* (Geneva, IFPMA, 1986).

21. *Impact of Genetic Engineering Regulation on the West German Biotechnology Industry* (Dielheim, Federal Republic of Germany: RauCon Biotechnology Consultants GmbH, 1989).

22. "New European Release Rules Ratified" (news item), *Nature* 344 (1990): 371.

23. F. E. Young and H. I. Miller, "Deliberate Releases in Europe: Overregulation May Be the Biggest Threat of All," *Gene* 75 (1989): 1–2.

24. "UW Researchers Stymied by Genetic Test Limits," *Capital Times* (Madison, Wisc.), 16 March 1988, p. 31.

25. M. Ratner, "Survey and Opinions: Barriers to Field-Testing Genetically Modified Organisms," *Bio/Technology* 8 (1990): 196–98.

26. "Clouds Gather over the Biotech Industry," *Wall Street Journal*, 30 January 1989.

27. "The Antiscientific Method," *Wall Street Journal*, 29 October 1987, p. 11.

13

Regulatory Law

Richard B. Stewart

The law's response to the environmental, health, and safety risks of genetically engineered products and processes reflects society's ambitions and aspirations, but also its fears and doubts.[1] Genetic engineering based on selective breeding is an ancient practice. Techniques of genetic engineering at the cellular level were developed during the past century. Recombinant DNA and other new techniques developed during the past two decades have greatly enhanced the power and precision of genetic engineering by enabling genetic selection and transfer to be performed at the molecular level. The old and new techniques, collected under the label *biotechnology*, offer great health, environmental, and economic benefits. But they also potentially present three categories of adverse effects.

The most direct, first-order risks are potential adverse impacts of bioengineered organisms on humans, plants, animals, and ecological systems through pathogenic injury or environmental disruption and displacement. For example, a bioengineered organism introduced in order to assist the growth of a crop on a given plot might migrate and harm natural ecosystems or other crops on adjacent land. Broader, second-order risks consist of adverse economic, social, and political transformations that may result from the application of powerful new agricultural or human genetic applications. For example, bioengineered crops might alter economies of scale, favoring large farms over small farms or vice versa. These transformations may in turn lead to indirect, third-order environmental effects. The development of bioengineered herbicide-resistant crops might encourage greater use of chemical herbicides, for example, causing ecological damage, or biotechnology might encourage excessive reliance on a few crop strains, reducing genetic diversity and increasing the vulnerability of agriculture to blights.

The most nebulous but potentially most profound third-order effects are changes in our conceptions of ourselves and of nature resulting from the wide-scale exercise of new and powerful techniques to alter the characteristics of humans and other organisms. Some fear that the enhanced powers conferred by new genetic engineering techniques might result in widespread displacement of existing species by manmade organisms, undermining our conceptions of wilderness and nature and fueling human hubris. Fears that human genetic applications of molecular engineering will be abused or will destroy our understanding of human individuality can engender hostility to all applications of molecular engineering.

The law's present response to biotechnology may seem paradoxical. Organisms created by the new molecular techniques are regulated far more stringently than organisms produced by the older techniques, even though there is no evidence that the risks posed by the new group are greater. Moreover, genetically engineered products, which thus far have not caused any known damage to human health or the environment, are regulated more stringently than documented hazards such as air pollution.

Several factors explain this paradox. Cultural attitudes and institutional forces tend to produce more restrictive legal controls on new technologies than on established ones.[2] Also, legal controls on technology respond not only to scientific evidence but also to popular perceptions of risk. Public attitudes toward the environmental, health, and safety risks of molecular engineering are ambivalent and often fearful. Such attitudes may be explained in part by a risk-averse stance toward the uncertainty posed by any new technology. Because molecular engineering is such a new technology, both the problem of uncertainty and the law's difficulty in coping with it are especially great. Public attitudes about the risks of molecular engineering are also influenced by concern over the second- and third-order consequences of its widespread use. Although such consequences are not the stated object of the regulatory laws dealing with biotechnology, which are limited to first-order environmental, health, and safety risks, these broader concerns exert a powerful gravitational influence on the law's operation.

Legal Regulation of Technology

The American legal system uses three basic techniques to deal with the environmental, health, and safety risks of new technologies. Tort liability plays a general backup role in deterring unduly risky conduct and in compensating injury. The dominant strategy, however, is to identify and control potential hazards in advance through a combination of two other techniques: risk assessment and regulation. These measures are carried out in the first instance by administrative agencies, subject to review in the courts.

Risk assessment and regulation can be understood simply as means of maximizing the net social benefits of new technologies by identifying the risks posed by their deployment and eliminating those risks whose costs exceed their benefits. Some advocates of technology assessment, however, believe that it should play a broader role. The law should, in their view, reflect broader societal values and serve as a forum for public debate and education about the evolving relation between those values and science and technology.[3]

In the United States the courts play a far more prominent role in overseeing administrative assessment and control of technological risks than in other countries. American judges not only make sure that administrators obey the Constitution and statutes but also review administrators' exercise of discretion. This practice reflects the nation's traditional distrust of the competence and impartiality of government bureaucracies, its commitment to a system of separated powers, and its faith in the courts as guardians of important societal values. Administrative agencies can easily claim greater technical and managerial competence than judges. But formal procedures and judicial review are thought to be justified in order to promote more informed and careful agency decisions and to ensure that these decisions respond to broader societal concerns and values.

Legal control of biotechnology raises institutional questions similar to those encountered in the regulation of other technologies, including fossil fuel combustion, vaccination, information processing, nuclear energy, and synthetic organic chemistry. For example, how well equipped are governmental institutions to do timely and accurate risk assessment? How can regulatory risk management best identify and control unwarranted hazards while avoiding excessive regulatory delay and overkill? How should scientific uncertainty or controversy be dealt with? How can courts review administrative decisions that turn on complex scientific and technical issues? Who should have the right to invoke the courts' reviewing powers, and how extensive should those powers be?

Legal regulation of biotechnology, however, differs from regulation of other technologies in at least three important respects. First, molecular engineering is a new technology that has developed since the deployment of extensive environmental, health, and safety regulations. Most regulatory programs represent belated efforts to catch up with the untoward consequences of established technologies. Because genetic engineering at the molecular level is the first major new technology to have been developed in the environmental regulatory era, it is an institutional test case of whether regulatory law has learned from past experience and has developed the capacity to anticipate, and to deal appropriately with, new types of risks. The fact that molecular engineering is still in a fledgling state creates special problems for risk assessment and legal regulation. While substantial knowledge has been accumulated about the effects of established technologies, in the case of products created by the new molecular techniques such knowledge is far more limited.

Second, the fledgling state of molecular engineering means that its regulation is necessarily directed at the "front end" of the research and development cycle. Controls on molecular engineering are targeted on laboratory research and field tests rather than on full-scale industrial and commercial applications. In American society, government may not control knowledge; it may only regulate its harmful applications. With a new technology, however, experimental applications posing possible but uncertain risks are themselves a vital source of new knowledge. As a result, regulation of molecular engineering has impinged more sharply on basic scientific inquiry than has regulation of older, better-understood technologies.

A third distinguishing factor is the deep concern and anxiety stirred by biotechnology. An eco-catastrophe on the scale of Chernobyl or Bhopal seems improbable. But genetic engineering often raises more fears than chemical or even nuclear technologies because its critics believe that it empowers human beings to "play God" by tampering with life itself, makes possible an "irreversible attack" on the natural order, and paves the way for a brave new world of environmental manipulation and totalitarian eugenics. These fears have led Jeremy Rifkin and other ideological activists to use litigation to obstruct the development of biotechnology.

In the absence of extensive experience with the effects of molecularly engineered organisms, these fears are likely to exert a greater effect on institutional decisions than is seen with established technologies. Law generally works best when differences of interest or value can be reduced to well-defined factual issues that are capable of being resolved through the adversary process. Because many disputes about the risks

associated with the new molecular techniques do not have this character, the law may be ill adapted to assess and manage biotechnology.
Americans may face choices between undertaking a pattern of experiments whose possible risks are not well defined, renouncing the development of a new technology and the advances in knowledge generated
by its application, and delaying and discouraging its development by invoking legal procedures as a ritualized means of assuaging social concerns. If so, technology assessment through law may fail at the very
point where it is most needed.

Administrative Regulation of Biotechnology

Scientists themselves were the first to call public attention to the
potential hazards of recombinant DNA techniques. In 1974 a group of
molecular biologists appealed for a voluntary moratorium on potentially hazardous recombinant DNA experiments. They also called for
the development of policies to control these hazards while allowing recombinant DNA research to go forward. In 1975 a conference of scientists voted to establish voluntary guidelines for improving containment
procedures in laboratories and limiting experiments involving pathogens or cancer-causing viruses.[4] These recommendations formed the
basis for guidelines adopted in 1976 by the National Institutes of Health
(NIH). The guidelines prohibited certain experiments involving recombinant DNA molecules, mandated the use of special laboratory equipment and containment measures with other experiments, required the
use of enfeebled vectors and host strains, and prohibited the planned
introduction of bioengineered organisms outside the laboratory.

As anxiety about genetic engineering rose, a 1977 meeting at the
National Academy of Sciences was interrupted by chants from activists
of "We will not be cloned."[5] Nevertheless, a year later the NIH lifted the
ban on certain experiments and progressively relaxed controls until
around 90 percent of laboratory work became exempt.[6] One reason was
the demonstration that a weakened form of *Escherichia coli* used in laboratories does not thrive in the human intestinal tract and "cannot be
converted to a harmful strain even after known virulence factors were
transferred to it using standard genetic techniques."[7] No demonstrable
hazards resulted from the rapidly expanding use of molecular engineering techniques in the laboratory. In 1978 the NIH amended the guidelines to allow planned introductions of bioengineered organisms outside the laboratory as well, on a case-by-case basis following review and
approval by the NIH.

Opinion among scientists on the guidelines is divided between
those who see statesmanship in calming the public and those who see

panic at imaginary fears. "Better safe than sorry" is counterposed to "Nothing ventured, nothing gained."[8] Some critics contend that as a result of the influence of microbiologists, the guidelines give insufficient attention to the concerns of ecologists. Others believe that the guidelines are either unnecessary or justified only as a politically expedient response to unwarranted public fears. The consensus, however, is that the guidelines and their modifications have been an appropriate response to concern about the safety of recombinant DNA laboratory research.

The NIH guidelines are legally binding only on recipients of federal research funds. Fear of possible tort liability, however, provides industry with strong incentives to follow the guidelines.[9] But the NIH is not a regulatory agency, and its monitoring and enforcement capacities are limited. It has required recipients of federal funds to establish institutional biosafety committees, composed in part of members outside the recipient institution, to take over the monitoring and enforcement task. Some observers are skeptical about the efficacy of this form of self-regulation.

As molecular engineering technology has progressed, attention has shifted from laboratory experiments to field trials and to commercial applications of bioengineered products. Concern has shifted from human health hazards to the risks of environmental disruption. The federal government has dealt with bioengineered products and processes through existing regulatory agencies and programs originally created to deal with industrial chemical, pharmaceutical, and agricultural materials and products. As administered, these programs generally require that an institution proposing to field-test or market a molecularly engineered product submit information to the appropriate administrative authority and obtain advance clearance on a case-by-case basis.

Plants and their products and pests are regulated by the Department of Agriculture under the Federal Plant Pest Act, the Plant Quarantine Act, and the Noxious Weed Act. Agriculture also regulates veterinary biologics under the Virus-Serum Toxin Act and claims authority over genetically engineered animals based on the Food Security Act and other statutes. Human drugs and biologics, medical devices, foods, food and color additives, and animal drugs produced by biotechnology are regulated by the Food and Drug Administration (FDA) under the Food, Drug, and Cosmetic Act and the Public Health Service Act. Pesticides are regulated by the Environmental Protection Agency (EPA) and the Animal and Plant Inspection Service in the Department of Agriculture under the Federal Insecticide, Fungicide, and Rodenticide Act. Both microorganisms (other than pesticides) and industrial chemical products created by biotechnology are subject to regulation by the EPA under the Toxic Substances Control Act. Wastes from biotechnological fermentation processes are subject to EPA regulation under the statutes dealing with air and water pollution and toxic wastes. The Occupational Safety and Health Administration in the Department of Labor regulates workplace hazards.[10]

These regulatory statutes are based on the character or use of a product or the type of residuals created by a process, rather than the technology used to create the product or process. The structure is therefore consistent with the views of many microbiologists and other scientists that organisms developed by molecular engineering should not be regulated on a basis different from that for "natural" organisms or for those developed by older techniques. In practice, however, some agencies have regulated molecularly engineered organisms more stringently than other organisms. This approach follows that of the NIH guidelines, which set an influential precedent for the stricter regulation of organisms developed by molecular techniques.

For example, the Department of Agriculture has imposed special regulatory requirements under the Plant Pest Act on organisms that are genetically engineered.[11] In pesticide regulation, the EPA has eliminated the exception for field tests and now requires advance administrative approval for experimental field applications of genetically engineered microbiological pesticides.[12] Under the Toxic Substances Control Act, the EPA requires that premanufacturing notices be submitted for intergeneric microorganisms, which are generally produced by molecular techniques, and denies them the regulatory exemption for research uses of "small quantities" of chemicals.[13] The EPA is currently considering regulating as pesticides genetically engineered plants with pest-resistant properties, even though it has never regulated the use of naturally occurring plants with such properties.[14] The law's administration is thus inconsistent with the emerging scientific consensus that an organism's mode of origin alone is not probative of the environmental, health, and safety risks that it may pose.

Despite this emerging consensus, differences of view within the scientific community over the possible hazards of planned introductions of bioengineered organisms are likely to persist. Most molecular biologists tend to discount the hazards presented by planned introductions, while many ecologists believe that the impact of engineered organisms on complex natural systems is poorly understood at present and that releases may pose appreciable risks of environmental disruption through such mechanisms as colonization, genetic transfer, and the pathogenic effects of pest controls on nontarget species.

The effort to deal with genetically engineered products and processes under the existing regulatory statutes has created problems. In some cases there is a legal question whether regulatory statutes drafted for other purposes apply to bioengineered products. For example, the Toxic Substances Control Act gives the EPA regulatory authority over a "chemical substance," defined to include, "any organic or inorganic substance of a particular molecular identity," including "any combination of such substances occurring in whole or in part as a result of a chemical reaction or occurring in nature." There is substantial agreement that this language applies to recombinant DNA itself, but there is doubt about bioengineered organisms.[15] The biotechnology industry has not yet mounted a legal challenge to the EPA's claim that the statute gives it regulatory authority over genetically engineered microorganisms, apparently out of fear that a court decision rejecting such authority could provoke Congress to enact even more restrictive controls. There is also doubt whether the FDA has authority to regulate new organisms that are beneficial to crop plants but potentially harmful to the surrounding environment.

In other instances the problem seems to be duplication of authority. For example, microorganisms that may be plant pests or animal pathogens are potentially subject to regulation by both the EPA and the Department of Agriculture under a variety of statutes. Approval of bovine interferon was long delayed by a regulatory dispute over whether it was a "veterinary biologic" under the jurisdiction of the United States Department of Agriculture (USDA) or a "new animal drug" under the jurisdiction of the FDA.

Regulatory agency staff are often neither trained nor experienced in the relevant scientific disciplines. The current patchwork regulatory structure also impedes data gathering and development of risk assessment methodologies on issues that cut across jurisdictional lines. Different regulatory agencies applying different statutes assess and deal with the same or similar risks in different ways.

The FDA, which is concerned with the human health benefits of bioengineering, has approved a substantial number of genetically engineered drugs and vaccines and has permitted hundreds more to begin clinical trials. The EPA, which views its mission almost exclusively in terms of controlling hazard, has been substantially more conservative. It has cleared a few field tests but insists on extensive and deliberate review procedures. For example, when Monsanto proposed to conduct field tests of two microbes altered with a toxin to control root worms in corn, the EPA Science Advisory Board recommended approval of the tests, despite Monsanto's failure to present all of the ecological testing required by EPA regulations, because the data already submitted by Monsanto were thought to be adequate to justify a conclusion that the risks were minimal. The EPA, however, decided to require further testing, including submission by Monsanto of new test protocols, and EPA approval of those protocols, before testing began. This decision was assertedly an important factor in Monsanto's decision to cut back its efforts in agricultural applications of biotechnology. This deliberate pace reflects its fear of unfavorable publicity and of political and legal challenges from biotechnology opponents, as well as the greater uncertainties involved in field releases.

The Department of Agriculture, which views the promotion of agriculture as its primary mission, follows a less formal and more relaxed regulatory style. This style may be appropriate, since plants are thought to pose fewer environmental hazards than bioengineered microorganisms, which are primarily regulated by the EPA. Critics, however, complain that the USDA's mission of promoting agriculture has compromised its commitment to regulation and that its informal procedures make public and judicial review more difficult.[16]

These problems have generated legislative proposals to create a new, unified regulatory structure for biotechnology-based products and processes.[17] The proposals have thus far made little headway. Federal administrators, who are inclined to favor incremental change and fear a possible loss of authority, maintain that existing statutory arrangements are adequate. The federal executive and industry have opposed new legislation, fearing that it would invite opposition to biotechnology and result in more restrictive regulation. In order to respond to inadequacies in the existing structure, an interagency coordinating committee was established in 1986. Its aim has been to develop more consistent policies among existing statutory programs, but the interagency jurisdictional conflicts and gaps have not been fully resolved.

NEPA and Litigation

In the United States the National Environmental Policy Act (NEPA), as interpreted and enforced by the federal courts, has become the focus of the approach to technology assessment through legal procedures. Decisions by the federal government to fund or grant regulatory approval to a project or activity are potentially subject to NEPA. If the decision represents a "major federal action significantly affecting the quality of the human environment," the responsible federal agency must prepare a full *environmental impact statement*. In situations where a proposed action may have such an effect, the agency must first prepare an informal *environmental assessment* in order to determine whether a statement is required.

If an environmental impact statement is required, the responsible agency must assemble and analyze information about the potential environmental effects of the proposed action and of alternatives, including a decision not to fund or approve the project. This information and analysis is then incorporated into a draft statement that is made publicly available. The public is invited to submit written comments, and one or more informal public hearings are held. The agency must then prepare a final statement that responds to the public comments and must take them into account in deciding whether to fund or approve the project. This process generally takes a minimum of a year, or longer if the issues are complex. The costs of the statement's preparation and revision, in money, time, and staff resources, both to federal agencies and to project sponsors who seek federal regulatory approval, are often substantial.[18]

Litigants who wish to challenge an agency's regulatory decisions or its compliance with NEPA can obtain review in the federal courts if they can establish both that they are suffering or may suffer "injury in fact" from the agency's action and that they fall within the "zone of interests" protected or regulated by a relevant statute and thus have legal standing. This "standing" requirement has generally been liberally interpreted by the courts. For example, an environmental group has standing to challenge an assertedly inadequate administrative regulation if one of its members is exposed to increased risk as a result of the adoption of the regulation.

In reviewing decisions under a particular regulatory statute, such as the Toxic Substances Control Act, the courts determine whether the agency acted in accordance with its statutory authority, whether it followed the procedures required by statute, and whether the agency's exercise was a "reasoned" one with adequate factual and analytical support in the administrative record. In cases based on NEPA, the courts have said that the act's requirements are "essentially procedural," designed to inform the agency, to make public the information and policy considerations involved, and to allow the public a role in the decision process. In NEPA cases, courts review whether the potential environmental effects of a proposed action are sufficient to require the preparation of either an environmental assessment or an environmental impact statement. When either of these documents has been prepared, the courts ask whether it adequately considered the effects of the proposed action and of alternatives.

Litigants often challenge the scientific and technical bases for agency decisions, contending either that the decision was factually unsupported or that the agency violated NEPA by failing to prepare an environmental impact statement. On such questions, courts give substantial deference to agencies with the knowledge and judgment that specialization affords. Judge David Bazelon went so far as to assert that "substantive review of mathematical and scientific evidence by technically illiterate judges is dangerously unreliable."[19] Moreover, retrial of all issues in the courts would be wasteful and demoralizing.

Democratic premises may lead courts to defer to agency policy choices, on the ground that agencies are more subject to political control and accountability than judges are. But judicial deference cannot be total. If agencies were free to "find" whatever facts they wished to fit a particular result, judicial review would be reduced to a mere feint. The very enactment of NEPA reflects a congressional judgment that scientific and technical determinations implicate larger social values, and that legal requirements are needed to ensure that administrators adequately consider these values. The fear was that mission-oriented agencies would slight environmental values. Some considerations thus favor judicial deference to agency findings. On the other hand, others counsel strict judicial scrutiny. Dealing with the horns of this dilemma is a continuing challenge for the courts.

Nearly all of the court actions brought against biotechnology regulatory decisions have been based on NEPA and initiated by the Foundation on Economic Trends and its founder, Jeremy Rifkin.[20] Rifkin believes that genetic engineering is immoral. The foundation is the organization that he has created to generate public opposition to biotechnology and to use litigation to delay or stop it.

This organization's first and most successful case was *Foundation on Economic Trends v. Heckler*, which challenged NIH approval of a proposed experiment by the University of California involving field releases of bioengineered Ice⁻ bacteria on crop plants. The experiment would have been the first field release of a bioengineered microorganism in the United States. The NIH had prepared an environmental assessment of the experiment and concluded that a full environmental impact statement was unnecessary because the environmental risks presented were insignificant. In 1984 a federal District Court enjoined the experiment, finding that the NIH had violated NEPA and that it must prepare an environmental impact statement, or at least a more adequate environmental assessment.[21] In 1985 the Court of Appeals affirmed.[22]

The courts found that the record did not contain adequate consideration of the dangers should the test bacteria disperse from the test site and cause environmental disruption. The only written material in the administrative record reflecting consideration of this matter was a statement that the number of recombinant DNA−containing organisms would be "small" and subject to "processes limiting survival." This was held to be insufficient.[23]

The holding was criticized by some scientists on the ground that the bioengineered bacteria in question were closely similar to naturally occurring bacteria and presented no danger of environmental disruption. The very fact that a seemingly unaffected party like Rifkin was given legal standing to challenge the experiment in court was criticized for allowing "responsible scientific activity" to be held "hostage to legal interference by zealots."[24]

The courts' decisions in *Heckler*, however, were based not only on the Ice⁻ experiment in question but also on what they viewed as a pattern of NIH failure to address adequately the environmental hazards of planned introductions of bioengineered organisms. When the NIH had in 1978 amended the original guidelines to permit planned introductions on a case-by-case basis, the discussion of their hazards was brief. Between 1978 and approval of the Ice⁻ experiment in 1983, the NIH had approved three other federally funded field release experiments, which for other reasons were never held. At no point did the NIH articulate criteria to evaluate the risks associated with planned introductions and distinguish the acceptable from the unacceptable. The courts viewed the NIH decisions as ad hoc judgments that had failed to yield a consistent regulatory policy or provide a reasoned basis for the exercise of the agency's discretion.[25]

It would have been difficult to compile any more information about the microbial agent without conducting the field test. The judges wanted the NIH to articulate the decisional principles to be followed in such a circumstance, which could include consideration of "the hidden costs of saying no."[26] The social costs associated with not conducting the experiment included the economic and ecological benefits that prompt development of new, bioengineered products and processes might provide, as well as the advances in scientific understanding that the data generated by the experiment might spur.

The courts' decisions and the criticisms that they drew from scientists reflect differences in the legal and scientific cultures. Consensus among informed experts may often be an adequate basis for a policy decision within the scientific community. Courts, however, are reluctant to rubber-stamp such a consensus when the decision in question raises significant social concerns that Congress has recognized in a relevant statute such as NEPA. At the same time, judges have great difficulty in directly evaluating experts' judgments about technical issues. They often resolve this dilemma by emphasizing procedure, that is, requiring administrators to write down their decisional criteria and to follow them in agency opinions explaining why the experts' judgment was an appropriate basis for an administrative decision made in the face of adverse criticism.

Since *Heckler*, the Foundation on Economic Trends and Rifkin have lost nearly every legal challenge that they have brought involving federal regulation of biotechnology. All of these challenges were based on NEPA; some involved asserted violations of particular regulatory statutes as well. They failed to stop the Department of Agriculture's "animal productivity research" program.[27] They were unsuccessful in challenging establishment of the interagency coordinating committee and in seeking to force the EPA to impose financial responsibility requirements on persons field-testing bioengineered pesticides.[28] And they failed to overturn the EPA's approval of a later Ice⁻ field experiment similar to the one involved in *Heckler*.[29] Their only post-*Heckler* victory, against the construction of a new biological warfare facility by the Department of Defense, was a distinctly limited one. They had challenged the department's environmental assessment as inadequate because it failed to discuss possible experiments at the facility involving genetic engineering. The court found the assessment deficient on quite different grounds.[30]

This decline in Rifkin's fortunes in court reflects several factors. In response to *Heckler*, regulatory agencies such as the EPA have adopted elaborate procedures for processing applications and have written formal opinions explaining their decisions.[31] There are also signs of a shift in judicial attitudes. In three of the later cases, the courts denied both the foundation and Rifkin standing to sue because the plaintiffs had failed to establish a discrete and immediate physical risk to which they would be exposed from the challenged action, as opposed to a generalized concern over biotechnology.[32] Finally, judges appear to have become somewhat more willing to defer to administrative decisions to allow experimentation to proceed.[33] This willingness may reflect judicial concern about both the social costs of imposing undue legal burdens on biotechnology and the potential for abuse when ideologues systematically use litigation to block all development of a given technology. Similar concern appears to underlie Supreme Court decisions limiting the use of environmental laws by antinuclear groups.[34]

Future Issues

Rifkin's failures do not mean that agencies regulating biotechnology will enjoy immunity from future legal challenges. As more decisions on field releases and commercial applications are made, not only litigants concerned about biohazards but also the biotechnology industry will find fresh grounds for legal challenge. Through careful legal planning the agencies should be able to avoid or win most NEPA challenges by preparing adequate environmental assessments or, if necessary, environmental impact statements.

Future litigation is likely to focus on the agencies' exercise of authority under particular regulatory statutes. NEPA requires that agencies undertake a general assessment of the environmental consequences. The implementing regulatory statutes raise discrete issues, including the extent of regulators' authority or obligation to impose controls; the extent of controls that may be imposed; and risk assessment and other criteria. Approval or disapproval of experiments and products, and the extent to which regulators must follow trial-type or other adversarial procedures in making decisions, will raise further issues of administrative law that courts may be asked to resolve. The current maze of regulators' statutes will no doubt breed lawsuits.

Courts are unlikely to interpret substantive regulatory statutes in ways that will prove major impediments to the development of biotechnology. Nevertheless, the threat of litigation from the public will induce conservatism on the part of regulators, particularly at the EPA. Applicants for regulatory approval will have to submit extensive information and endure elaborate decision-making procedures, creating uncertainty, delay, and expense that will slow the delivery of economic and environmental benefits from genetic engineering. As a fledgling technology, genetic engineering will be particularly vulnerable to regulatory law's tendency to impose excessive burdens on new products and processes.[35]

As the focus of litigation shifts from environmental analyses to specific issues of regulatory authority and control, the dilemma of judicial review of complex scientific issues will recur. The dilemma is especially acute in the case of new technologies like molecular engineering, where experience with applications is limited: more reliance must be placed on judgment, and in developing principles to deal with risk the policy element is greater because of increased uncertainty about the nature and extent of such risk. In addition, molecular engineering raises special moral concerns in the public eye, which in turn may tend to influence judges' and administrators' legal and scientific judgments.

The future of judicial oversight of administrative regulation of biotechnology may well replay the classic debate over the broader subject of court review of the decisions of expert agencies. Because of his concern about the technical ignorance of judges, Judge Bazelon advocated an essentially procedural role for the courts:

> What courts and judges can do . . . and do well when conscious of their role and limitations . . . is scrutinize and monitor the decisionmaking process to make sure that it is thorough, complete, and rational; that all relevant information has been considered; and that insofar as possible, those who will be affected by a decision have had an opportunity to participate in it . . . By articulat-

ing both their factual determinations and their value preferences
. . . administrators make possible effective professional peer review, as well as legislative and public oversight.[36]

Judge Harold Leventhal pointed out, however, that even this "procedural" role inescapably requires judges to make substantive judgments, since in order to determine whether a scientific or technical
issue is sufficiently important and controverted to warrant full consideration and whether that consideration is adequate, courts must exercise some responsibility for judging the underlying merits:

> The substantive review of administrative action is modest, but it
> cannot be carried out in a vacuum of understanding. Better no
> judicial review at all than a charade that gives the imprimatur
> without the substance of judicial confirmation that the agency
> is not acting unreasonably. Once the presumption of regularity in
> agency action is challenged with a factual submission, and even
> to determine whether such a challenge has been made, the
> agency's reasoning and record has to be looked at . . . On issues
> of substantive review . . . the judges must act with restraint. Re-
> straint, yes, abdication, no.[37]

The question of judicial competence to assess scientific issues thus
cannot be escaped through purely "procedural" solutions. Particularly
when confronted by new technologies, judges face the dilemma of trying to determine whether agencies have assessed the evidence and the
residual uncertainty in a reasonable fashion when critics argue that
the risks are far greater than the agency has recognized.[38] This conundrum raises larger issues about the relation between science and regulatory law.

The development of biotechnology will greatly expand the number
of experiments and products subject to regulation, placing a serious
strain on the existing case-by-case systems of administrative clearance.
This problem is likely to be particularly acute at the EPA, which has
already experienced severe administrative problems in regulating pesticides and industrial chemicals not based on biotechnology. The pressure of applications threatens either to create an enormous backlog or to
make most regulatory approvals perfunctory. The most likely response
is to embrace a "wholesale" approach to regulation by adopting requirements for categories of experiments, developing standardized risk
assessment procedures, and concentrating regulatory resources on the
most serious risks.

Some categories of organisms may pose greater risks, or greater uncertainty about risks, than others. Large-scale applications pose greater

risks than field tests. Adjusting the degree of regulatory scrutiny to the potential risk presented is the most promising means of reducing regulatory overkill. The EPA has already taken steps to develop different reporting and review requirements for different types of planned introductions, based on the potential risk that they present.[39] Such a system of regulation can also give priority to the development of organisms that provide new, environmentally benign ways of combating pests, delivering nutrients to plants, and cleaning up pollution. This step could help encourage mainstream environmental groups to be active supporters of biotechnology. One such group, the Environmental Defense Fund, has publicly supported an application by Crop Genetics for regulatory approval of a bacterial protein into corn seeds to make the seeds lethal to the corn borer.

However, the effort to calibrate regulatory requirements with the likely extent of environmental risks and benefits posed by different categories of organisms will require advances in risk assessment capabilities, particularly those concerning the ecological impacts of planned introductions. It will therefore highlight characteristic differences in views between microbiologists and ecologists.

Development of the biotechnology industry will also place increasing pressures on regulatory monitoring and enforcement. Several instances of unauthorized field experiments involving bioengineered microorganisms have already occurred. It is very difficult, however, for government to detect and control small-scale field releases if experimenters do not voluntarily disclose their activities. As genetic engineering technology becomes more easily accessible and as commercial pressure to accelerate product development grows, the danger of unauthorized experiments will increase. Moreover, a time-consuming and burdensome federal regulatory process encourages industry and researchers to take shortcuts with requirements that they view as unjustified.[40] The interagency coordinating committee created under the federal regulatory framework has proposed the creation of environmental biosafety committees, on the model of institutional biosafety committees, to take on the monitoring task and also to relieve the regulatory backlog created by centralized Washington review of every decision. Many proponents and also opponents of vigorous regulation, however, view this proposal with distrust.

The scientific view that mode of origin is not determinative of risk should, as a logical matter, direct greater regulatory attention to organisms developed by traditional forms of biotechnology, which until now have been subject to little or no control. Because such techniques are widespread, however, there will be severe difficulties in creating a regulatory system that targets those uses associated with significant risks.

Courts may also be reluctant to read the Federal Insecticide, Fungicide, and Rodenticide Act and the Toxic Substances Control Act as extending to plants and other organisms produced by selective breeding and other traditional techniques. There is no logical basis for treating these organisms differently from ones produced by molecular techniques. But the powerful social tendency to impose stricter scrutiny and control over new technologies may prevail over logic.

The development of biotechnology, the growth of litigation on questions of the substantive regulatory authority of agencies, and interagency turf battles will sooner or later result in legislation to amend the statutes now used in regulating biotechnology and to resolve ambiguities and conflicts in their application to bioengineered products. In addition, as commercial use of bioengineered products and organisms proliferates, the tort system may come to play an increasingly important role in shaping biotechnology research and applications.[41]

The international dimensions of biotechnology and its regulation will also assume increasing importance.[42] The fact that biotechnology is widely seen by nations as a key to future economic growth and international competitiveness is a complicating factor. On the one hand, concerns about international competitiveness will moderate the stringency of domestic regulation. On the other hand, conflicts among national regulations, including testing requirements, will burden international trade and investment in biotechnology.[43] These burdens will be especially heavy if national governments use environmental, health, and safety regulations to shield their domestic biotechnology industry from international competition. The efforts of the European Economic Community to exclude U.S. beef products made from cattle fed growth hormone, is an example.

Another prospect is disparity in the stringency of regulatory requirements between developed and developing nations. There are already instances where U.S.-based researchers have conducted field tests of bioengineered organisms in other nations with less stringent regulation. This practice may impede dissemination and use of the results of such tests, because regulatory and scientific authorities in these nations often lack the capacity to monitor experiments closely and because businesses conducting the experiments may have commercial and other incentives not to publicize the results. Multinational companies may increasingly use developing nations to conduct tests, or to market bioengineered products that would be barred by developed nations; the issues would be similar to those now presented by overseas testing of new drugs and by international trade in pesticides.

Conclusion

The response of the legal system to the new technologies based on genetic engineering has in general been one of profound conservatism in regulatory law. Preexisting statutes and judicial practices have been applied without considering the special problems and opportunities created by a fledgling technology, so the similarities to regulation of earlier technologies are far more apparent than the differences. As in other areas of environmental regulation, the strategy is basically reactive: innovation is regarded as a private sector responsibility, and regulators have a limited, gatekeeping role. No agency or authority deals primarily with genetic engineering or seeks to guide its overall development.

The nascent character of biotechnology, however, imposes special strains on established regulatory practices. The relatively undeveloped state of risk assessment and the limited experience with planned introductions of bioengineered organisms make it difficult for regulators to articulate and consistently apply decisional criteria in ways that satisfy the ideal of the rule of law voiced by the courts in *Heckler*. Opponents of biotechnology seize on the inevitable uncertainties of present methods to demand that field tests and applications be postponed until some unspecified future time when a methodology of "predictive ecology" will have somehow been developed by other means.

The potential risks of any new technology cannot be fully known in advance. But because biotechnology relies on redesign of living systems, its inevitable uncertainty leads to especially acute public anxiety. Moreover, because experience with field applications of genetically engineered organisms is so limited and because ideology flourishes when facts are scarce, public perceptions are likely to play a greater role, and expert judgment a lesser one, than in regulatory decisions involving established technologies. Regulators subject to potential challenge in the courts, and also Congress, are likely to react by insisting on additional studies, additional procedures, and a risk-averse approach to decisions. Similarly, in the absence of well-framed factual issues that can be resolved through litigation, courts are likely to respond to acute public concerns by requiring administrators to follow additional procedural steps.

These responses are likely to impose an especially heavy burden on genetic engineering technologies that promise great social and environmental benefits. When hard facts are scarce, ethical judgments about what to do in the face of uncertainty come into play. But when hard facts are scarce, it is also vital to draw on informed scientific judgment in order to gauge as accurately as possible the degree and character of uncertainty regarding the probability and extent of possible harm. It is therefore especially important to find new ways of drawing on this kind of judgment in making regulatory decisions about biotechnology. The opportunities for doing so exist in three regulatory settings: administrative, judicial, and monitory or enforcement.

First, at the federal regulatory agency level, greater visibility and prestige should be provided for scientific advisory boards. This step would promote public and judicial confidence in regulatory decisions based on their determinations. The NIH guidelines cannot be transported wholesale to traditional regulatory agencies. But such agencies could strengthen their scientific base by relying on organizations such as the National Academy of Sciences, the National Science Foundation, and the NIH to create panels on which agencies could draw to staff advisory boards.

Second, these same panels could serve as a pool that judges could draw on in appointing expert witnesses, special masters, or advisory groups. These disinterested science advisers could help educate judges about scientific and technological issues and could moderate the extreme positions taken by experts hired by the parties. This practice, which has proved successful in West Germany, would be far superior to "Science Court" proposals, which would have the scientific issues presented in litigation referred to a panel of scientists who would decide the issue after presentations by advocates to the parties in the litigation. Such proposals confuse the very different adversary methods of science and law and overestimate the possibility of separating facts and values.[44]

Third, at the monitoring and enforcement level, panels of scientists might be asked to staff biosafety review committees to help ensure that experiments are consistent with regulatory requirements. It is awkward to call on scientists to take responsibility for steps that could curtail the pursuit of knowledge. But given the dysfunctions of centralized regulation, refusal by the scientific community to play a role could exacerbate the danger of overly intrusive or obstructive controls.

A necessary complement to these steps is a more sustained effort by the scientific community to educate the public about genetic engineering. If forced to choose between government regulators who are reluctant to interfere and critics who find the regulation of biohazard inadequate, the public is inclined to believe the critics. In a survey a large majority of those interviewed (63 percent as against 28 percent) were inclined to believe environmental groups over government agencies in disputes over biohazard. However, 58 percent believed that unjustified fears have seriously impeded valuable medical applications. The scientific community needs to amplify its voice in the debate, although there are many problems in doing so effectively.

If scientists take an increased role in regulatory policy, they may help relieve the danger of undue obstacles to the pursuit of knowledge.

The use of nonpartisan science panels rather than adversary experts should enhance the influence of scientific judgments in regulatory decisions, and also the deference accorded by courts to these judgments. These panels should ease the task of regulators in evaluating uncertainty, and in developing principles for dealing with it that the courts will respect. But this approach also raises problems. As is noted in several of the chapters in this volume, the scientific community is itself divided on the key issue of the risks of planned introductions. Most molecular biologists tend to believe that molecular engineering is unlikely to pose serious hazards, while many ecologists are convinced that the threat of environmental disruption is significant and real.

The scientific community faces additional problems. The prominent involvement of some scientists in commercial applications of genetic engineering may dilute the respect normally accorded to science by the public. And as commercial applications of molecular engineering grow, the ability of scientists to influence decisions will also decline, while that of business will increase.[45] Involvement in public policy is an unaccustomed and uncomfortable role for many scientists, who may rightly fear that they will be led to compromise their best scientific judgments in favor of positions that respond to public preconceptions and are therefore politically more palatable. As Allan Campbell pointed out earlier in this volume, some scientists who participated in development of the NIH guidelines now regret their role, fearing that it gave credibility to regulatory steps that were not justified and that have come to do more damage than good.

These problems may tempt some scientists to reassert the autonomy of science, rather than being parties to its regulation. They might, for example, assert that biotechnological experimentation should be immune from government regulation unless strong evidence of serious hazard is adduced. Such a position might be backed up by appeals to the First Amendment and to freedom of scientific inquiry.

Indeed, there is a danger that excessively stringent regulation of molecularly engineered products and processes may be imposed as a result of public perceptions of risk that are not justified by the scientific evidence. In the American political system such perceptions are likely to influence legislative, administrative, and even judicial decisions. It would be a mistake, however, to fight regulatory overkill by appealing to constitutional principle. Such an appeal will not succeed in courts of law and will likely fare no better in the court of public opinion. Opponents of biotechnology can raise rival principles, such as the need to preserve human integrity and the natural order against greedy meddling. Such principles may have far greater popular appeal than freedom of scientific inquiry.

Scientific communications undoubtedly enjoy First Amendment protection, although their exact status in reference to political speech and other forms of protected expression has never been judicially defined. But it by no means follows that scientific experiments enjoy constitutional protection against environmental, health, and safety regulation by the government.[46] The law distinguishes speech from action 'that goes beyond speech, especially when that action threatens physical harm to others or to the environment.[47] To be sure, scientific experimentation is a necessary precondition of scientific knowledge and hence of scientific speech. It may be argued that since speech is protected, so should the experiments on which it is based. But such a broad reading of the First Amendment, which would extend constitutional protection to all manner of dangerous and unlawful activity, has never been accepted by the courts.[48] Environmental health and safety regulation of biotechnology experiments is constitutionally valid so long as there is some reasonable basis for supposing a risk of harm.

The case might be otherwise if regulation of biotechnology were based on an ideological fundamentalism that held biotechnology to be immoral or sacrilegious. The First Amendment prohibits the suppression of ideas because they are believed to be false or dangerous; ordinarily, government has no power to restrict expression because of its subject matter or its message. Publication of the results of biotechnology experiments could not be prohibited on the ground that it would alter peoples' conceptions of themselves or the natural order. The Framers had faith in reason and the progress of science and knowledge. They did not fear the Copernican revolution. They would have no greater terror of Darwin or the double helix.

It follows that if biotechnological experimentation were prohibited on ideological grounds, with the aim of blocking the creation of knowledge that would have a pernicious effect on human perceptions and values, such prohibition would be constitutionally invalid. But it would be practically impossible to establish judicially that the real purpose of regulatory legislation was ideological, so long as a permissible alternative ground of regulation, such as potential environmental, health, and safety effects, were plausibly invoked.

Despite popular impressions based on dramatic constitutional cases, the general instinct of American law is to avoid grand appeals to abstract principles such as freedom of scientific inquiry or preservation of human integrity. As is illustrated by the saga of nuclear energy as well as by biotechnology, both opponents and proponents of new technologies are often driven by ideology. The law, however, tries to sidestep ideology in favor of concrete concerns that can be tested by forensic methods of fact finding. For example, litigants generally cannot challenge a regulatory decision unless they can show that it threatens to cause them some particularized physical or economic injury. For example, the Supreme Court has held that NEPA does not require an environmental impact statement on the restart of one of the reactors at Three Mile Island because of asserted psychological fears by residents in the region.[49] This principle has been invoked by the courts to deny standing to sue in a number of cases.

It may be claimed that even so, the courts provide too much opportunity for opponents of science to distort and delay, and that standing to sue should be tightened still further. This claim overlooks the potential utility of court litigation as a means of moderating popular controversy that cannot be easily contained by politics alone. Moreover, it ignores the fact that the goose's sauce is also the gander's. Under existing law, the government must show particularized justification for infringing on citizens' liberty or welfare, including their liberty to engage in scientific inquiry. When the justification consists of environmental or health risks, the leeway for government controls necessarily becomes somewhat greater. But the courts continue to demand a concrete showing of need for regulatory intrusions.[50] Proponents of biotechnology should not be eager to abandon the courts' fact-based protection against unjustified regulation in favor of broad ideological principles.

The liberal state attempts to maintain neutrality among competing conceptions of the good by focusing on pragmatic considerations. Existing biotechnology regulation, which has dealt almost entirely with the broadest first-order physical effects, is firmly within this tradition. Such an approach limits the demands made on the courts and relegates the adjustment of competing social visions to more general political and social processes. By repudiating protection of ideology as such, it encourages a secular political culture in which scientific inquiry and learning can flourish. Ideology often remains the motive force behind litigation, whether to expand or to contract the stringency of regulation. But its influence is confined by the law's general focus on concrete issues.

The best antidotes to overregulation are first, the advance of scientific understanding, and second, the improvement in transmission of that understanding, through such means as scientific panels, to government decision makers and to the public. These are not fully satisfactory solutions. Overregulation based on misapprehension may block or delay the experiments necessary to advance scientific understanding. More effective transmission of scientific understanding to appropriate audiences is difficult. But no better remedy seems available.

The future development and use of molecular engineering also raise large questions that transcend the physical risks on which current regulation focuses. The enhanced power of these techniques has deepened concerns by some groups over the "industrialization" of natural ecosystems through bioengineered forests, aquafarms, and food crops.[51] Others fear that the use of molecular techniques to modify the genetic makeup of organisms, including human beings, may cause unwelcome changes in our perceptions and values.[52] In a juridical democracy, these issues will inevitably be considered in some form by the courts. The present applications of biotechnology, however, are too limited to make it useful to engage such grand issues.

The paradox of technology assessment through law thus remains. The nascent phase of a technology involves too much forensic uncertainty to allow the fact-oriented instincts of the law to engage successfully the far-reaching social, economic, and environmental changes that the development of the technology may spawn. Responding to this uncertainty by insisting on protracted procedures is more of an evasion than a solution, and it may deter valuable innovations. At the same time, if a nascent technology, surviving the initial legal gauntlet becomes too firmly entrenched, the law may not be able to deal later with unforeseen systemic adverse effects.

Notes

The views expressed in this chapter are my own and do not necessarily represent the views of the Department of Justice, where I am an assistant attorney general, Land and Natural Resources Division.

1. This chapter deals only with environmental, health, and safety regulation of bioengineering processes and products, not with the other legal aspects of biotechnology, including human genetic applications, intellectual property issues, and military applications.

2. See P. Huber, "Safety and the Second Best: The Hazards of Public Risk Management in the Courts," *Colum. L. Rev.* 85 (1985): 227.

3. See L. Tribe, "Legal Frameworks for the Assessment and Control of Technology," *Minerva* 9 (1971): 243.

4. S. Olson, *Biotechnology: An Industry Comes of Age* (Washington, D.C.: National Academy Press), pp. 64–65.

5. F. Press et al., Preface to Olson, *Biotechnology*, p. iii.

6. Olson, *Biotechnology*, p. 66.

7. Office of Technology Assessment, *Impacts of Applied Genetics: Micro-organisms, Plants, and Animals* (Washington, D.C.: U.S. Government Printing Office, 1981), p. 201.

8. Olson, *Biotechnology*, pp. 65–66.

9. Ibid.

10. In addition, a number of municipalities and states have adopted or are considering the adoption of ordinances and statutes regulating biotechnology.

Most existing laws deal with DNA laboratory experiments, generally requiring all such experiments to comply with the NIH guidelines. Despite initial industry fears, such regulation has not proved a substantial impediment. Proposals have been made, however, for regulation of planned introductions of bioengineered organisms in the field. Local opposition to field tests may become an important factor in the future.

11. Department of Agriculture, Introduction of Organisms and Products Altered or Produced Through Genetic Engineering Which Are Plant Pests or Which There Is Reason to Believe Are Plant Pests, 7 C.F.R. §340 (1989).

12. Environmental Protection Agency, Statement of Policy: Microbial Products Subject to the Federal Insecticide, Fungicide, and Rodenticide Act and the Toxic Substances Control Act, 51 Fed. Reg. 23,313 (26 June 1986).

13. Ibid.

14. See Notice, Pesticide Regulations for Pesticidal Transgeneric Plants, 54 Fed. Reg. 17,269 (24 April 1989).

15. See Note, "Living Organisms as Chemical Substances: The EPA's Biotechnology Policy under the Toxic Substances Control Act," *Rutgers Computer & Technology L. J.* 13 (1987): 409; Note, "The Rutabaga That Ate Pittsburgh: Federal Regulation of Free Release Biotechnology," *Va. L. Rev.* 72 (1986): 1529.

16. See U.S. General Accounting Office, *Report on the U.S. Department of Agriculture's Biotechnology Research Efforts* (Washington, D.C.: GAO, 1985).

17. See G. Jaffe, "Inadequacies in the Federal Regulation of Biotechnology," *Harv. Envtl. L. Rev.* 11 (1987): 491.

18. Private organizations and state government agencies seeking approval of the federal agencies must in many cases finance and carry out some of the preparation tasks for the environmental impact statement themselves, including the development and analysis of data. The federal agency, however, must draft the actual statement and scrutinize any data furnished by the applicant. Certain federal agencies, pursuant to published regulations, require permit applicants to reimburse the agency for the costs of completing the statement. See, for example, *Sohio Transp. Co. v. United States*, 766 F.2d 499 (Fed. Cir. 1985); *Nevada Power Co. v. Watt*, 711 F.2d 913, 928–31 (10th Cir. 1983).

19. *Ethyl Corp. v. EPA*, 541 F.2d 1, 67 (D.C. Cir.) *(en banc)* (Bazelon, C.J., concurring), *cert. denied*, 426 U.S. 941 (1976).

20. In 1978 a different plaintiff unsuccessfully sought to enjoin recombinant DNA laboratory experiments at the NIH. Since the NIH had done an environmental impact statement on the experiment, the court found that it had fully complied with NEPA and its own guidelines. *Mack v. Califano*, 447 F. Supp. 668 (D.D.C. 1978).

21. 587 F. Supp. 753 (D.D.C. 1984).

22. 756 F.2d 143 (D.C. Cir. 1985). The District Court had also enjoined the NIH from approving any other field release experiments. The Court of Appeals vacated this aspect of the District Court's order as overbroad.

23. 756 F.2d at 153–55. The Court of Appeals stated that the only discussion on the issue of dispersion was the view of an evaluator that "although some movement of bacteria toward sites near treatment locations by insect or aerial

transport *is possible,* the numbers of viable cells transported has been shown to be very small; and these cells are subject to biological and physical processes limiting survival." 756 F.2d at 153. In the court's view, this discussion was inadequate because it failed to address the possible environmental impacts of dispersion if it occurred. A more thorough discussion by the NIH of the reasons why it believed that no harm would result even if substantial transport occurred might have satisfied the court.

24.   B. Davis, "Judge Sirica Chills Genetic Research," *Wall Street Journal,* 13 July 1984.

25.   The District Court concluded that "the 'standard' for granting a waiver can only be described as whatever it takes to win the confidence of . . . at least a majority of the RAC [Recombinant DNA Advisory Committee] and the subsequent approval of the Director of NIH." 587 F. Supp. at 760. It also found that the "Director refused to articulate any standards, requirements, or values which would guide the future exercise of his discretion to permit deliberate release experiments." Ibid. at 762. The Court of Appeals agreed that "NIH has not yet displayed the rigorous attention to environmental concerns demanded by law." 756 F.2d at 146.

26.   F. J. Dyson, "On the Hidden Costs of Saying 'No,'" *Bull. Atom. Sci.* 31 (June 1975): 23.

27.   *FET v. Lyng,* 817 F.2d 882 (D.C. Cir. 1987).

28.   *FET v. Thomas,* 661 F. Supp. 713 (D.D.C. 1986).

29.   637 F. Supp. 25 (D.D.C. 1986).

30.   *FET v. Weinberger,* 610 F. Supp. 829 (D.D.C. 1985).

31.   See, for example, the EPA procedures and explanations involved in *Thomas,* 637 F. Supp. at 26–29.

32.   In the cases where the foundation and Rifkin challenged an entire research program or general regulations, as opposed to a specific experiment, the courts also held that the case did not present a controversy sufficiently focused to be "ripe" for judicial review.

33.   See S. Deatherage, "Scientific Uncertainty in Regulating Deliberate Release of Genetically Engineered Organisms: Substantive Judicial Review and Institutional Alternatives," *Harv. Envtl. L. Rev.* 11 (1987): 203, 223–28 (criticizing *Thomas* for excessive deference to administrators).

34.   See *Vermont Yankee Nuclear Power Corp. v. NRDC,* 435 U.S. 519 (1978); *Baltimore Gas & Electric Co. v. NRDC,* 462 U.S. 87 (1983).

35.   See R. Stewart, "Regulation, Innovation, and Administrative Law: A Conceptual Framework," *Calif. L. Rev.* 69 (1981): 1256; Huber, "Safety and the Second Best." Representatives of the biotechnology industry already complain that it "is being regulated into submission." W. Brill, "Why Engineered Organisms Are Safe," *Issues in Science and Technology* 4 (1988): 44. See also B. Davis, "Paranoia Invades Report on Man-Made Microbes," *Wall Street Journal,* 16 June 1988.

36.   D. Bazelon, "Coping with Technology through the Legal Process," *Cornell L. Rev.* 62 (1977): 817, 823.

37. *Ethyl Corp. v. EPA*, 541 F.2d 1, 68–69 (D.C. Cir.) *(en banc)* (Leventhal, J., concurring), *cert. denied*, 426 U.S. 941 (1976).

38. The dilemma cannot be resolved satisfactorily by simply assigning the burden of judicial proof entirely to one party or the other. In the face of substantial uncertainty, such an assignment would be controlling in all cases. A more modulated approach is required.

39. See Note, "Federal Regulation of the Biotechnology Industry: The Need to Prepare Environmental Impact Statements for Deliberate Release Experiments," *Santa Clara L. Rev.* 27 (1987): 567, 577–79.

40. The failure to obtain EPA approval of an open-air experiment was cited by one professor as justification for a form of "civil disobedience" against "almost ludicrous" regulatory requirements. See G. Strobel, "Tree Scientist Tests Bacteria, Disobeying U.S. Regulations," *New York Times*, 14 August 1987, p. A1, col. 1; p. A12, col. 3. Some scientists have only recently come to appreciate the burdens imposed by federal regulation that businesspeople have bemoaned for decades.

41. See Note, "Designer Genes That Don't Fit: A Tort Regime for Commercial Releases of Genetic Engineering Products," *Harv. L. Rev.* 100 (1987): 1086.

42. See R. Stewart and M. Martinez, "International Aspects of Biotechnology: Implications for Environmental Law and Policy," *Oxford J. Envtl. L.* 2 (1989): 158.

43. See M. Bechtel and A. Cayla, *Etude internationale comparative des réglementations concernant les biotechnologies* (Paris: AFNOR, 1987); G. D. Sakura, *Review and Analyses of International Biotechnology Regulations* (Cambridge, Mass.: Arthur D. Little, 1986); J. Gibbs, *Biotechnology and the Environment: International Regulation* (New York: Stockton Press, 1987).

44. See J. Martin, "The Proposed 'Science Court,'" *Mich. L. Rev.* 75 (1977): 1058. For an examination of other approaches to accommodating law and science in the regulation of risk, see J. Yellin, "High Technology and the Courts: Nuclear Power and the Need for Institutional Reform," *Harv. L. Rev.* 94 (1981): 489; J. Yellin, "Science, Technology, and Administrative Government: Institutional Design for Environmental Decisionmaking," *Yale L. J.* 92 (1983): 1300.

45. See E. Savitz and E. Wyatt, "Fulfilling Their Promise: Wondrous Products and Even Profits Are in Sight for Biotech Firms," *Barron's*, 25 September 1989, p. 6, col. 1.

46. See G. Robertson, "The Scientist's Right to Research: A Constitutional Analysis," *So. Cal. L. Rev.* 51 (1977): 1203; R. Delgado and D. Millen, "God, Galileo, and Government: Towards Constitutional Protection for Scientific Inquiry," *Wash. L. Rev.* 53 (1978): 349; J. Ferguson, "Scientific Inquiry and the First Amendment," *Cornell L. Rev.* 64 (1979): 639.

47. Even the communication of scientific information may be subject to legal constraint in appropriate circumstances, such as publication of classified information on construction of the hydrogen bomb. See *United States v. The Progressive, Inc.*, 467 F. Supp. 990 (W.D. Wis. 1979), appeal dismissed, 610 F.2d 819 (7th Cir. 1979) (mem.).

48.   See S. Carter, "The Bellman, the Snark, and the Biohazard Debate," *Yale Law and Policy Rev.* 3 (1985): 358.

49.   *Metropolitan Edison Co. v. People Against Nuclear Energy,* 460 U.S. 766 (1983).

50.   See *Industrial Union Department AFL-CIO v. American Petroleum Institute,* 448 U.S. 607 (1980) (invalidating OSHA standard limiting occupational exposure to benzene, because of lack of evidence of safety benefits).

51.   See "Biotechnologists Forging New Links in Food Chain," *Washington Post,* 25 September 1989, p. A3, col. 1

52.   See generally L. Tribe, "Technology Assessment and the Fourth Discontinuity: The Limits of Instrumental Rationality," *So. Cal. L. Rev.* 46 (1973): 617.

14

Summary and Comments: The Scientific Chapters

Bernard D. Davis

Although the preceding chapters stand on their own, it seems useful to summarize such a heterogeneous collection, to highlight areas of disagreement, and to add some further relevant materials. I shall concentrate here on the scientific contributions, and in the next chapter Harvey Brooks and Rollin B. Johnson will comment on those concerned primarily with public policy. Dr. Brooks brings to this task the perspective of a long involvement in parallel problems encountered by the physical sciences and technologies. Though these two concluding chapters are having the last word, where we disagree with other contributors the reader is obviously the final judge.

Chapter 2: The Background

After having laid out in Chapter 1 the main issues discussed in this book, I review in Chapter 2 the development of genetics from its classical to its molecular stage, including the role of bacterial genetics—a field that looms large in current controversies. This presentation not only outlines the foundations of current studies, but it also introduces technical vocabulary that will be useful for the subsequent chapters.

The chapter also speculates on some broad conceptual contributions of molecular genetics, including an answer to an ancient, fundamental question: What distinguishes living from nonliving matter? It turns out that the unique features in the organization of living matter do not involve new physical forces or principles, as some physicists had expected. They lie instead in the storage and transmission of information in giant molecules.

There are two major classes of information, stored in two kinds of informational molecules. Hereditary information is stored in the sequences of the four different nucleotides in nucleic acids. Acquired information about the environment is transmitted to appropriate targets by a special class of proteins, called allosteric: molecules that undergo transient, reversible changes in shape upon interacting with specific chemical or other stimuli. These allosteric shifts were first recognized in bacteria, as economical responses to changes in the supply of various nutrients. However, in a remarkable example of the unpredictable consequences of fundamental discoveries, allostery has turned out to be the key to all responses of cells to environmental stimuli, including the function of hormones and the nervous system in higher organisms.

The discovery of two classes of informational molecules also has philosophical implications, because it provides a physical basis for the two interacting kinds of knowledge postulated in Kant's epistemology. That acquired by experience is stored in the brain, through changes in allosteric proteins. The a priori knowledge that we are born with is programed in our genes and expressed in the development of the nervous system, resulting mostly in patterns of response shared with all members of the species, but also exhibiting fine individual differences.

The chapter ends with speculations about other classical questions in biology that have been brought to the fore again by molecular genetics, such as the concepts of reductionism and purpose.

Chapter 3: Microbes

Allan Campbell discusses the regulation of genetically engineered bacteria. Since I have also been deeply concerned with this problem, I shall review this chapter, and the closely related following one, at some length.

Campbell was involved in the development of the National Institutes of Health recombinant DNA "guidelines," and he presents from the inside an exceptionally candid and thoughtful critique of this history. He notes that political activism in the universities in the preceding years had been pressing scientists to take moral responsibility for controlling the uses of their discoveries. This atmosphere contributed to an exaggeration of the dangers from the new recombinants, and the molecular geneticists who raised the alarm soon recognized their error. However, their initial view, in which all engineered organisms are guilty until proved innocent, still dominates the discussion. Campbell blames these scientists even more than the activists or the politicians, because they used the authority of their science to fortify positions that were not scientifically warranted.

I would note, perhaps uncharitably, a possible additional, unconscious reason for the initial exaggeration of the dangers: the greater the novelty and the threat of the new recombinants, the more powerful the impact of the discoverers who had made them possible. Certainly the early discussions of molecular recombination had a Promethean quality.

However, not all agree today that the initial reaction was exaggerated. Recognizing that various earlier technologies have generated unforeseen costs, Simon Levin and Margaret Mellon, in later chapters, argue that in matters of public policy scientists do not have the privilege of restricting themselves to firmly based scientific considerations. Accordingly, in their view, the unprecedented power of the new biology initially required, and still requires, that we take even remote hypothetical scenarios seriously.

The charge is not illogical. On the other hand, not only have the fifteen years of expanding work with recombinants failed to produce a trace of harm, but at the start there was already a large body of knowledge that contradicted the original reasons for expecting inadvertent harm from recombinants. Since the initial predictions still influence public perceptions, it is important to try to understand why they were taken so seriously by a group of brilliant scientists.

First, the molecular biologists who created recombinant DNA were understandably proud of their powerful, highly reductionist experimental approach, for it had succeeded in solving all kinds of problems that had seemed insuperable to earlier biologists. It was therefore natural for them to approach the problem of hypothetical risks in the same way. But while molecular biologists had created that problem, its assessment really did not lie in their field; assessment was a problem in bacterial pathogenesis, in epidemiology (the science concerned with understanding the spread of disease), and in evolutionary biology.[1] Unfortunately, the scientists who led the discussion of the hazards did not appreciate that the principles already established in these fields would, in the end, be more reliable guides than any specific experiments. Moreover, at the Asilomar conference in 1975, which launched the public discussion of the problem, few representatives of these fields were present; and in the subsequent debate few felt close enough to molecular genetics to be willing to take a strong position.

The central focus was therefore not on assessing the risk. Instead, the risk was assumed to be real, and the challenge was to eliminate it by clever experimentation—that is, developing attenuated organisms ("biological containment") to supplement the physical containment usual for dangerous microorganisms. Hence, for several years work was permitted only with these enfeebled laboratory strains. And when the guidelines were subsequently relaxed the public explanation still emphasized experimentation, that is, the results of tests on the limited survival of certain recombinants under various circumstances. I suspect, however, that the main reason for the relaxation was simply increased familiarity with harmless recombinants.

Campbell notes several consequences of the decision to search for solutions through experiments rather than by building on established knowledge. First, there was a long delay in recognizing that regulations should logically be based on the nature of the organisms, and not on the technique used to produce them—product rather than process. Second, when the research was extended to planned introduction to the environment, most of the earlier debate over hypothetical dangers from laboratory experimentation had to be repeated. Finally, the apparently prudent approach, compromising between those who saw great danger and those who saw very little, continues to prevail.

Aaron Wildavsky, an iconoclastic political scientist, emphasizes in a later chapter that the justification for this approach was more political than scientific. He therefore originally entitled his chapter "Goldilocks Is Wrong." For if there is significant danger that novel recombinants could initiate an irreversible, catastrophic spread, we should not allow any testing in the environment. On the other hand, a low risk, like that from similar, familiar bacteria, should not require special treatment of the novel organisms. The middle-ground solution—Goldilocks's "just right"—is therefore not prudence; it is an approach that lacks a logical basis.

Among the pertinent scientific considerations, Campbell notes several evolutionary principles. Perhaps the most important one is that the microbes present in the environment saturate their existing ecological niches. Engineered competitors therefore do not enter a new, open field: an extraordinarily rich variety of organisms is already present, each adapted to using a specific, limited set of foods, and in order to displace them the invaders would have to be better adapted. (Illustrating how sparse is our knowledge of bacterial ecology, recent analyses of the sequence specificity of nucleic acids extractable from some specimens of soil indicate that over 99 percent of this material comes from unknown bacterial cells, not detected by present methods.)

Accordingly, shifts in a microbial population ordinarily result from an environmentally induced alteration in the distribution of organisms already present, rather than from the introduction of novel organisms.

Lakes become smelly because they accumulate a high concentration of nutrients, which allows the microbial population to become dense enough to exhaust the oxygen; the noxious organisms that thrive in the absence of oxygen then take over. Even if we wished to engineer organisms with a selective advantage, in most circumstances we would not know how.

There is one exception, however, noted in Levin's chapter: when we are dealing not with the existing environment but with one that we have altered in a known way, we can often predict corresponding genetic changes that would improve an organism's survival and proliferation. For example, in an environment containing an antibiotic a gene for resistance to that inhibitor would be advantageous.

Another early cause of apprehension was unpredictable novelty. However, Campbell points out that recombination between distant organisms is not so novel after all: the plasmids and viruses that are used in the laboratory to transfer genes between different bacteria (and higher organisms) carry out the same "horizontal" interspecies transfers, at a very low rate, in nature. Indeed, it appears that all the DNA in the world forms a continuous pool in which fragments move occasionally from one organism to another—even between bacteria and mammals. The vast majority of the resulting recombinations, however, are not found in the world today, because they were at a selective disadvantage and hence died out.

A third argument, which I have emphasized more than Campbell, concerns the importance of the scale of the introduction of an engineered bacterium. Much of the worry over deliberate introduction is built on the model of toxic chemicals, where the danger is indeed roughly proportional to the amount released. But microorganisms are different: they can multiply but they can also die. Accordingly, if an organism is competitive it could multiply and spread from a small escape as well as from a large introduction. If it is not competitive it is bound, by definition, to die out even after a large introduction. From this perspective, the reasons for having relaxed the regulations over laboratory experiments should carry much weight in considering the later problem of deliberate introductions.

A fourth consideration is that to pose a risk, a novel microbe must meet *both* of two conditions: it should be able to spread, and it should do harm if it does spread. Campbell considers mostly the first, but he also discusses briefly the issue of harm, which pretty much boils down to causing an infectious disease. (In some discussions ecologists have expressed concern about other possible harms to the environment, but these seem rather vague.) Here a principle of evolutionary biology is again important: success as a pathogen, like successful adaptation to any ecological niche, requires a harmonious balance among all the in-teracting genes in the organism's makeup. Having a mutation with a powerful effect, such as production of a potent toxin, does not itself make an effective pathogen. In fact, only the rarest of mutations im-prove, rather than weaken, an organism's genomic balance, whether as a pathogen or in some other role; and the scale of our mutations and selections in the laboratory is puny compared with that of evolution in nature.

Finally, I would like to comment on another argument, not touched on by Campbell: the assumption that even if the chances of inadvertent harm from recombinants are exceedingly small, the potential catastro-phe is large, so we must take no chances. The Goldilocks rebuttal to this argument was noted above. But a more concrete rebuttal is the like-lihood of intermediate steps. Pathogens vary widely in the severity of their effects on the host, and also in their ability to be transmitted to new hosts; multiple genes influence each of these traits; and for either trait, as was noted in the chapter by Henry Miller, low degrees are much more common than a severe threat. Hence, if a mutation or a recom-bination did inadvertently endow a harmless organism with patho-genicity, it would be much more likely to cause a mild disease than a catastrophic epidemic. The mild illness would then give us early warn-ing that further changes might lead to a serious pathogen—and no such mild illnesses have been detected. Although this argument has not been prominent in public discussion, I consider it a strong one.

Chapter 4: Ecology

Simon Levin, an ecologist, faces the difficult problem of discussing the conflict between his colleagues and microbiologists. Ecologists, dedicated to studying the interactions of organisms and the environ-ment, have a legitimate professional interest in the novel engineered microorganisms, and they have generally believed that their experience with visible plants and animals provided a necessary knowledge base for assessing the risks. Most microbiologists and molecular biologists, however, believe that the microbial world is so different from the world of visible organisms, in fundamental ways, that extensive experience with microbes, and with the associated field of epidemiology, provides a more reliable basis for judgment.

To be sure, there exists a field of microbial ecology. However, it is not sufficiently developed to provide a solid basis for risk assessment. And while the advent of genetic engineering clearly justifies expansion of this field, it does not so clearly justify holding up obviously safe ex-periments while ecologists are developing criteria for certification.

Levin was a member of the committee that prepared the first report of the National Academy of Sciences on engineered organisms in the environment, and so he was exposed more than most ecologists to the views of colleagues who did not share ecologists' apprehensions. I am encouraged by his confidence that the remaining differences are slight and will soon disappear; and I deeply respect his effort to reconcile views that have led to a most unusual split in the scientific community. At the same time I must spell out where I disagree with his arguments.

A major area of agreement is that Levin squarely favors product and not process as the basis for regulating organisms. Moreover, he repeatedly emphasizes the need to facilitate applications of engineered bacteria to environmental problems, and he is confident that classes of organisms justifying exemption can be identified. On the other hand, when he discusses the official report of the Ecological Society of America he does not criticize, as Campbell does, its insistence on continuing case-by-case evaluation. Instead, he notes that he himself is more sanguine about creating exempt classes: we simply have not yet developed the proper procedure.

Meanwhile, years have passed with unsatisfactory regulations, and the promised procedure still seems vague and elusive. One must wonder whether the lag arises mostly from the need for more research on proper safeguards, or, as Levin suggests early in his chapter, from the need of ecologists for time to become familiar with the new field. He begs for patience. But that is a great deal to expect of investigators or industries that have invested heavily in preparing for field tests for potentially valuable and admittedly safe organisms. Certainly, as Levin proposes, it is equally fallacious to assume either than all recombinants are dangerous or that none are. But the case-by-case approach, unwilling to accept any exempt classes, seems to build precisely on the former fallacy. Accordingly, until ecologists accept the principle of exemption of a large class of nonpathogenic organisms, most microbiologists are likely to doubt that there is as much agreement between the two groups as Levin sees.

Among the specific arguments, the model of imported organisms that have become pests was clearly the main basis for the initial concern of ecologists over engineered organisms. Since that model still has strong public appeal, I shall consider it here in some detail.

The basic fact is that many thousands of non-native animals and crop and ornamental plants have been transported between continents and have given profit or pleasure, but a few have multiplied beyond control and have caused serious economic and esthetic damage. Ecologists have every reason to be concerned to prevent further instances, and it is easy to understand their initial fear that some novel recombinants might recapitulate such problems. Logically, however, the analogy is weak: restoring altered organisms to the same environment is very different, in its probability of evolutionary success, from transplanting unchanged organisms into a new environment. What is at issue is not novelty per se but the adaptedness of the novel organism to the environment.

For example, the rabbit population in Europe is restrained, at a tolerable level, by interactions with various predators and parasites. In Australia these restraining forces were absent, and so imported rabbits multiplied explosively there. An engineered organism, in contrast, is not one already adapted by evolution and now given such a vacuum to fill. Instead, the organism has been changed; and as I emphasized earlier, this change will almost always decrease to some degree its adaptation to its environment. I am pleased to find Levin offering an additional basis for reassurance about engineered microbes: if they should displace indigenous species, they are unlikely to harm the ecosystem.

Levin dismisses another charge: that some ecologists have seen the issue of engineered organisms as an opportunity to increase support for work in their field. But it is not certain whether he is being realistic or is projecting his own high standards on others. For example, the Environmental Protection Agency has supported a good deal of research on differences in the rate of disappearance of various bacteria added to soil in the laboratory; but to me as a microbiologist this narrowly focused work seems unlikely to contribute much either to developing a useful basis for regulations or to advancing our fundamental knowledge of microbial ecology.

Finally, to fortify the argument that the scale of introduction is important, Levin brings in the neutral theory of evolution (a development that emerged from our new ability to identify mutations not only by phenotype but also by DNA sequence). For readers interested in this rather technical point, I shall add a dissenting comment.

The neutral theory has emerged as a supplement, and not an alternative, to the neo-Darwinian picture (see Chapter 2) of natural selection for favored variations. For example, many substitutions of one nucleotide for another in a gene do not change the functional properties, or often even the structure, of the product; hence, they are not selected either for or against. Such neutral mutations therefore readily accumulate in evolution. They are important for evolution because they broaden the base for building further mutations that may have functional significance. But mutations with advantageous phenotypic effects, and not neutral mutations, continue to provide the major force guiding the course of evolution.

Levin is formally correct in stating that according to the mathematical theory of neutral evolution the probability of a gene becoming fixed (i.e., attaining stable significant numbers) in a population is directly proportional to its initial frequency. But according to the same theory the probability is also *inversely* proportional to the initial size of that population. A more complete statement would therefore be that without selection the probability of fixation of a gene is proportional to the *ratio* of its frequency to that of its alternative forms in the population.

With bacteria, how should one define the denominator in that ratio? Because the world of soil bacteria is global, the figure must be huge. The quantities of an engineered organism that one could introduce would therefore be negligible, in terms of the ratio that is critical for neutral evolution.

An even stronger objection to invoking the theory of neutral evolution is that what concerns us about engineered organisms is not simply the possibility of spread; it is the possibility of a resulting deleterious effect on the environment. And a deleterious variation could hardly be neutral with respect to selection.

The argument from neutral mutations thus does not seem useful here. Competitiveness, and not the size of the introduced population, remains the decisive factor in the possible spread of an organism. If so, the difference between accidental escapes to the environment and planned introduction for research purposes is not important. Of course, a huge deposit of a noncompetitive recombinant on a field could have a local effect; but it would be transient.

However, I do agree with Levin and Campbell that commercial introduction, covering many square miles of land, will require more careful testing than field experiments. But the main reason would be the possibility of a large enough effect to cause local damage before the noncompetitive organisms died out; the concern would still not be the danger of forcing a spread. It is tempting to build on the model of infection, where a large enough dose of an ordinarily harmless bacterial strain may permit it to spread as a pathogen within some people (especially those with defective defenses); the key to this phenomenon is that the dose has exceeded the limited capacity of the individual's immune system. In the environment, in contrast, there is no basis for postulating such a threshold above which a noncompetitive organism will escape defenses and spread.

I fully agree with Levin on the importance of another kind of scale: spread of an organism not on its own but as a result of its expanding use in agriculture. The ecological effects of this process, a general problem, will be further discussed under the chapters on engineered plants and animals.

Chapter 5: Environmentalism

Ecologist Margaret Mellon works for the National Wildlife Federation, whose 5 million members are dedicated to preserving the aspects of nature of interest to hunters and fishers. The environmentalist movement that she vigorously defends is avowedly conservative, even about changes that promise to improve the environment—for not all earlier promises have been fulfilled. Such groups are surely useful to keep those who introduce changes alert and honest. But I have long wondered why the leading environmentalist groups have played such a prominent role in opposing, from the start, work with recombinant bacteria, which are so far removed form the visible world of nature. All biologists whom I know share with the environmentalist groups a strong feeling of concern for preserving species diversity; but the conflict over engineered organisms has introduced tensions.

I shall not comment on most of Mellon's arguments, whose central themes are suspicion of change, belief that new developments in technology are designed to enrich industry rather than to benefit the consumer, and confidence in a sustainable agriculture freed of dependence on added chemicals. Allan Campbell has summed up the position very well, in correspondence:

> The main point of disagreement is that people like Charles Arntzen and Franklin Loew seem confident that genetic engineering will in fact be used to meet the most important needs, through the normal workings of the free enterprise system. Others, like Margaret Mellon and myself, are less optimistic. The difference between her position and mine is that she would like to see the development of this and other technologies held hostage to substantial changes in social and economic policy, whereas I reject that course as both antithetical to freedom and counterproductive to the aims of human betterment.

I would note a perhaps even sharper difference: Mellon celebrates the increasing participation of the public in the debate, whereas scientists generally have reservations. In their view the public has a legitimate interest in policy decisions, and especially in those that involve value judgments; but in dealing with the technical problems of assessing risk the public is at a disadvantage and easily falls prey to demagogues. This issue will be taken up in Chapter 15.

Exhibiting what seems to me an excessive suspicion of industry, Mellon sees the use of bovine growth hormone, to increase milk production, as a source of economic benefit only to the manufacturer, and she takes a similar position on herbicide-resistant crops. (On the latter she offers an additional argument that has more substance: that use of herbicide resistance may increase the volume of applied herbicide.) The history of such Luddite opposition to more efficient technologies is a long one. Thus, the state of Wisconsin, in recently banning the use of growth hormone, is in effect repeating its interference decades ago with the distribution of margarine as a competitor to butter. However, the ultimate benefits to the consumer create an almost Darwinian pressure for increased efficiency and economy, and so opposition has only delayed and has rarely stopped such developments. This issue is discussed further in the chapters on agriculture.

Mellon also questions the assumption that productivity is synonymous with progress. And, indeed, the social dislocations and esthetic deterioration that may accompany technological advances are a cause of concern. But rather than deprive the consumer of benefits, and fight a losing battle against inevitable improvements in production, it would seem more useful to introduce economic measures to minimize the social costs. On a related issue, one can appreciate Mellon's sympathy for the countries whose sale of gum arabic or vanilla may be threatened by developments in genetic engineering. But it is hard to see how such developments, if economically successful, could be prevented, any more than the development of synthetic plastics to replace wood or metal.

Mellon also takes two scientific positions that I have already questioned: that large-scale introduction increases the chance of the irreversible establishment of a novel organism, and that the toxicity of chemicals is a good model for microbes, even though the latter self-destruct. On other, more general grounds I would also question the further conclusion that development of pollution-resistant plants is harmful because it would decrease motivation to remove pollution. (Do safety belts discourage careful driving?) A third scenario, the danger of future alterations in wild animals, hardly seems to merit serious concern.

Perhaps the most important recommendation in this chapter is that we need a single new agency to regulate environmental introduction of novel organisms. Our present approach uses existing legislation and distributes responsibility among the National Institutes of Health, the Food and Drug Administration, the Department of Agriculture, and the Environmental Protection Agency. The result is a hodgepodge, with wide variation in permissiveness and occasional uncertainty about jurisdiction. Logically there is much to be said for unification. However, because the public remains suspicious of engineered organisms, and because legislators tend to be responsive to public perceptions, many scientists fear that new legislation would be less adaptable to new developments than new regulation, and it might impair, rather than facilitate, the development of useful organisms.

Let me close with a further, discouraging conclusion: that many of the actions of environmentalists have been less reasoned and enlightened than the attitudes presented in Mellon's chapter. For example, it is hard to imagine a development that should please environmentalists more than replacing a chemical fertilizer by a modification of the bacterium Rhizobium, which provides nitrogen to plant roots. Yet a recent effort to test such an organism has been delayed for two years by uncertainty over jurisdiction, and by the objections of Mellon's federation.[2]

Chapter 7: Plants

Charles Arntzen, commenting on Mellon's chapter, addresses many of the questions that she raises. At the opposite end of the spectrum from her, he emphasizes—perhaps too confidently—the benefits achieved or expected in plant agriculture, including decreased use of chemicals. His fundamental point is that in a world facing a population crisis and increasing per capita food consumption, the conflict between agricultural productivity and the demands of environmental protection must be resolved in ways that do not significantly limit productivity. Like a physician disturbed by sentiments that impede the development of cures for ill patients, Arntzen is disturbed by obstacles to developments that aim at preventing worldwide starvation.

The bulk of Arntzen's chapter, a straightforward description of major advances in his field, does not require detailed comment. He relates how the traditional way to improve crop quality, by empirical selection of the best phenotypes, has been supplemented by several more refined genetic techniques: first hybrid seeds, then cell and tissue culture, and now implantation of genes for specific traits. In the last the genes can be introduced into a cell by a virus, a bacterial infection, an electrical field, or an explosion projecting microscopic metal particles coated with DNA.

Arntzen emphasizes that genetic advances are only one of the reasons for the dramatic increase in agricultural productivity in recent decades, and for the corresponding reduction in the fraction of our income required for food. Modern intensive agriculture is an integrated science, involving genetics, mechanization, irrigation, fertilizer, herbicides that are much cheaper than mechanical removal of weeds, and insecticides that greatly reduce crop destruction or damage. This "green revolution" has had a striking impact on underdeveloped as well as on technologically advanced countries, such as sevenfold increase in the yield of wheat per acre in Mexico, even before molecular genetics came onto the scene.

While the chapter concentrates on food production, it also notes some additional benefits from the genetic revolution. For example, the introduction of individual genes into transgenic plants is already making it easier to improve one trait without losing ground in another (e.g., trading shelf-life for flavor). In a more radical, likely future departure, the manufacture of valuable foreign proteins such as human hormones, now carried out mostly in recombinant bacteria, will probably become possible and much cheaper in transgenic plants.

Arntzen emphasizes the need for simplified procedures for small-scale field testing of engineered plants. I would suggest an additional reason for urgency: agriculture's lag, compared with medicine, in developing and applying basic scientific advances. Molecular genetic approaches have now attracted many brilliant scientists to plant biology. If this development is to fulfill its potential, safe tests of a novel transgenic plant in the field should not face any greater obstacles than work with modified bacteria or mice in the laboratory.

Chapter 8: Animals

Franklin Loew concentrates on genetic engineering in livestock. This research is now aimed mostly at improving productivity, but in the future it will increasingly be concerned with quality—for example, altering the composition of meat to provide more healthful nutrition, such as a lower level of saturation in the fat. Loew also notes briefly that genetic engineering has contributed to agriculture in other ways by providing hormones, vaccines, and diagnostic reagents manufactured in bacteria. In a novel further development, already under way, transgenic animals may replace bacteria as factors for this purpose. Thus, certain regulatory sequences activate a gene only in the mammary gland, so the resulting protein appears only in the milk; and purification is easier from milk than from bacterial cultures.

A very promising development is the cloning of exceptionally valuable animals from the separated cells of an early embryo. Each cell, inserted into the uterus of a foster mother, develops just like a fertilized egg. Since this procedure could in principle also be applied to humans, it is important to note that it yields unpredictable though identical genomes (just as with identical twins), because the initiating embryo has been a product of the randomizing process of sexual reproduction.

The term *cloning* also refers to a very different process: using the nucleus of a body cell, of an individual, already born to generate a predictable genetic copy. The idea of doing so with humans has understandably caused a good deal of apprehension. However, that possibility remains far too remote to warrant concern today. Indeed, there is suggestive evidence that in mammals, unlike plants or lower animals, the process of cell differentiation not only selectively turns various genes on or off in different kinds of cells, but also destroys their ability to code for a new individual—presumably as a result of irreversible small duplications or losses in the content of the genome. Cloning from adult mammals may therefore remain science fiction.

The patenting of engineered animals has been another cause of concern, but Loew points out that it is unlikely to lead to any radical change either in the nature of the animal world or in our relation to it. Though novel, this practice is not fundamentally different from such accepted practices as the registration of racehorses or the patenting of plant varieties; and like these activities, it will be restricted to strains of commercial value.

Another fear is that people may now use the power of genetic engineering to remake the living world by designing radically different animals. But even though we can change DNA at will, we cannot change organisms at will. There are severe limits to the changes compatible with viability, because an organism requires genomic balance—not only in the functioning individual, but also in the complex process of development. While advances in developmental biology will no doubt broaden the possible range of engineered alterations, there is no prospect, as far ahead as we can see, of converting any vertebrates into monsters (a vague concept), or into angels with four limbs plus wings.

In addition to agricultural animals, Loew discusses a class of cultivated animals especially important for medicine: mutants with defects that provide valuable models for the study of human disease. Genetic engineering is now greatly expanding this powerful approach by introducing custom-made defects. One example, a strain of mice with increased susceptibility to certain kinds of cancer, has proved valuable enough to be patented as a research tool.

From the perspective of a veterinarian concerned with the health of animals as well as with their use, Loew also discusses the increasing public concern for animal welfare and its impact on animal research. In the nineteenth century the antivivisection movement focused on species widely used as pets. But today, with greatly expanded medical research, with the increased recognition of problems created by science and technology, and with the public exposure of some investigators who have seriously neglected the welfare of their experimental animals, all animal research has attracted attention. The resulting public criticism has led to tighter regulation.

Virtually all scientists now accept the need for regulation. Nevertheless, there is widespread concern over reactions that are making animal research unreasonably expensive. For example, in some states the dogs used for experiments must be bred for the purpose, even though over ten times as many strays are put to death in pounds. Moreover, some people who press for ever greater comfort of the animals are really opposed to all animal research, and their demands are designed to discourage animal research altogether.

The greatest problem, however, is presented not by the animal welfare movement but by the animal rights movement. It is based on the claim of Peter Singer, a philosopher, that there is no logical basis for asserting the right of humans to exploit other species at all. He is technically correct, but the implications are not clear. For I would suggest that there is also no logical basis for asserting the contrary. Rights simply do not derive from logic.

Moreover, while our increased sensitivity to human rights has been a major advance in civilization, there are strong evolutionary arguments against extending the concept of rights to animals (though there are good reasons to prohibit cruelty). One consideration is that in the cycle of organic matter and the intricate ecological network produced by evolution, every species exploits others in some sense, and predator-prey relations contribute to stabilizing population densities; and in this evolutionary process humans have emerged, like many animals, as carnivores (though we can also survive as herbivores). Second, though our culture has developed several strong reasons to try to preserve each species from extinction (including an evolutionary perspective on the relatedness of all organisms, awareness of the irreversibility of any extinction, the esthetic pleasure many of us derive from the variety of the living world, and our sense of obligation to our heirs), such a respect for each species does not imply an obligation to try to preserve each member.

I would suggest that the animal rights movement can best be understood as an equivalent of religious fundamentalism, arising in a secular urban population—not exposed to the practicalities of farm life, feeling guilty over its own affluent, materialistic culture, and now searching for a transcendent, morally regenerating cause. And as in other fanatic movements, some members have been willing to employ violence against those who disagree. What is perhaps most surprising is that editors of some of our most serious magazines promote this movement.

The threat to medical research is large. For in the development of almost any new drug or procedure we must test in animals at some stage, unless we wish to transfer the risk to people. To be sure, for many purposes cultured cells and computers now replace animal tests. But the reason is not primarily that these alternatives spare animals: it is that they have become more informative and cheaper. Nevertheless, these tests have severe limits, and it is an illusion to expect them to replace animals completely. Biomedical research will be severely hampered unless society can balance a "humanitarian" attitude toward animals with the long-term humanitarian needs of patients.

Chapter 9: Medicine

In discussing the huge subject of molecular genetics and medicine, Theodore Friedmann, a clinical geneticist, concentrates on advances in diagnosing and understanding monogenic hereditary diseases. He describes first how we have come to recognize almost four thousand of these diseases, have identified the defective protein for a few hundred, and have developed therapies for a few. He also focuses on the implications of two monumental developments: the prospect of a complete description of the human genome sequence (the Human Genome Project), and insertion of corrective genes into humans (gene therapy).

Before discussing his material I shall list briefly some major impacts that molecular genetic techniques have already had on many other aspects of medicine—a topic that could easily have filled a large additional chapter. These advances include manufacture of medically valuable proteins such as insulin in bacteria; discovery of specific cancer-inducing genes (oncogenes) in tumor viruses; recognition that host cells have similar genes, which regulate normal cell multiplication but also mutate to cause cancer; identification and analysis of the receptors that bind regulatory molecules on various cells, and design of drugs to fit such receptors; progress in identifying factors that induce the differentiation of cells in embryonic development; identification of the molecules in an animal cell that determine its shape, specific adhesions, and fit in an organ; discovery of the complex gene rearrangements that give rise to the hundred thousand different cells and antibodies of the immune response; identification of the components responsible for the pathogenicity of various microbes, and use of these subunits to develop improved vaccines; analysis of the effects of many genes on cholesterol metabolism and hence on cardiovascular disease; and beginning genetic studies of the development and the function of the nervous system. In a word, the powerful tools of molecular genetics are becoming increasingly important in virtually every aspect of medical research.

Focusing on hereditary diseases, Friedmann describes in some detail two ways of locating their genes. In the traditional "forward" genetics the protein that is altered in a disease is first identified, such as the hemoglobin in sickle cell anemia. Analysis of its sequence leads, through the genetic code, to deduction of the corresponding nucleotide sequence, which can be synthesized. Hybridization of that molecule to a chromosome can be used to localize the gene, and comparison of se-

COMMENTS: THE SCIENTIFIC CHAPTERS 255

quences readily identifies the mutation in the abnormal allele. Synthesis of a short sequence containing the mutation then provides a direct probe that selectively hybridizes to a chromosome carrying the defect, thus making it easy to screen for the defect.

Another approach, called "reverse" genetics, starts by detecting some property that is present in the chromosomes of those in an afflicted family who carry the genetic defect but not those free of it. Analysis of the region with the difference can eventually identify the gene responsible for the disease. Moreover, the structure of the product of that gene may then suggest the molecular nature of the disease (as has been accomplished for cystic fibrosis).

The search for the site of the abnormal gene starts with its linkage to a site already known, located on a cloned fragment. The investigator then "walks along" the chromosome, by sequence analysis of contiguous fragments, to reach a gene that may be the one of interest. However, linkage was initially restricted to known genes, which are usually far apart; hence, the distance from known to unknown was often prohibitively long and the required complete set of contiguous fragments could not be identified.

The number of useful sites for linkage has now been greatly enhanced by the use of bacterial restriction endonucleases, which cut DNA at specific short sequences (restriction sites)—the same cuts on which molecular recombination is based (see Chapter 2). The locations of these sites vary widely among individuals (polymorphism), because mutations in them are often neutral. In recessive diseases, where the two parents each carry a defective allele of a gene on one chromosome and a normal allele on the homologous chromosome, a restriction site that is closely linked to the gene in one chromosome may be absent from the other. Hence, on treatment with the endonuclease the latter would yield the gene of interest on a longer fragment, cleaved at the next, unaltered restriction site in its sequence.

"Restriction fragment length polymorphism" (RFLP) employs a "library" of bacterial clones that together include all parts of a chromosome. Fragments of different length are readily separated by their different rates of movement in a gel, and hybridization with a labeled probe then identifies those carrying the gene of interest. When the normal and the abnormal alleles in a family are on fragments of different length, RFLP analysis readily reveals whether a child (or a fetus) has inherited zero, one, or two copies of the abnormal allele.

Another major use of restriction endonucleases is to map a genome. If two restriction enzymes that cleave at different sites are used to prepare two libraries, the fragments in each set will overlap in sequence with fragments in the other, and the recognition of these overlaps reveals the order of the fragments in the chromosome. This ordering produces a *physical map* of a chromosome, which defines distances in terms of number of nucleotides. This map is a much more reliable measure of distance than a *genetic linkage map*, based on the frequency of linkage of different genes in transmission to the next generation (Chapter 2). A high-resolution physical map, based on identifying restriction sites and other, unique short sequences in the DNA clones as signposts at short intervals, would greatly facilitate the localization of new genes, and it is a major object of the Human Genome Project.

Friedmann is enthusiastic about this project, which aims at producing a complete sequence of the human genome within fifteen years. However, many scientists have reservations. One concern is the impact of such a centralized, planned program on the style of biomedical research, whose success has rested so far on small-scale investigator-initiated projects. Another question is whether the attraction of legislators to the definite goal of complete sequencing, and the dramatization of the project, has not led to excessive funding compared with the rest of biomedical research.

The main reservation, however, is scientific: the importance of systematically sequencing the 3 billion nucleotides of the genome, equivalent to the letters in about ten sets of the *Encyclopaedia Britannica*. About 95 percent of this sequence is now provisionally thought to be "junk"—regions that give no hint of their possible function. In the normal course of science one would not undertake to sequence such a large unknown territory until studies on a sample had begun to lead to interpretations of function.

Indeed, the goal of complete sequencing was soon recognized to be prohibitively expensive with the available techniques, and it has been postponed. The present program is therefore focusing on improving methods of sequencing, and on more immediately attainable goals, such as a high-resolution map and sequencing regions of interest. But the goal of the complete sequence is still presented as justification for a crash program with a deadline. Supporters see it as one of the great challenges of our generation—the chemical underpinnings of human nature. But others see sequence as uninteresting until it can be understood, piece by piece, by correlation with function. For some years investigators have been establishing such links by proceeding from a messenger RNA, or a protein of known function, to its gene. The cost-effectiveness of the reverse approach, proceeding from blind sequencing to product and function, remains to be seen.

Moving on from the Human Genome Project, Friedmann next discusses gene therapy. Somatic cell therapy, introducing the needed normal gene into appropriate tissues in a person with a genetic defect, affects only the treated individual. Germ line therapy, correcting the defect in a germ cell, would reach all the cells of the future individual, and it would also be transmitted to future generations.

Somatic cell therapy is now generally considered morally acceptable, and it has been approved by the government and by several major religious organizations. The procedure uses a modified, avirulent retrovirus (or, in principle, other techniques) to introduce the gene into cells from a patient, which are then multiplied in culture and reintroduced into the patient. The first trials have started, using engineered bone marrow cells to try to correct a defect in the immune system.

It initially appeared that the loosely structured bone marrow might be the only tissue where the repaired cells could lodge correctly and replace defective cells. However, Friedmann notes that it may also become feasible for cells to be taken up by the liver or to be implanted in the skin. Nevertheless, for many hereditary diseases somatic gene therapy will be difficult because the abnormalities are present in multiple, highly organized tissues, such as secretory cells in the lungs and the pancreas in cystic fibrosis.

Germ line therapy would present much greater moral problems than somatic cell therapy. One objection is that it sets the dangerous precedent of interfering with human evolution. I would note that we already do so by the accepted practice of negative eugenics (i.e., genetic counseling to avert serious defects), since this activity slightly affects the composition of the gene pool of the next generation, in predictable ways.

Fortunately, we are unlikely to have to face the problems of germ line modification in the foreseeable future, for technical reasons.[3] First, the implanted DNA may enter the chromosome at a place where it will damage a needed gene—too great a risk in humans (though acceptable for creating transgenic animals). Second, prenatal diagnosis and elective abortion provide a much easier and cheaper solution, as long as the goal is to eliminate serious defects.

A more pressing social problem will arise if we are tempted to use gene implantation, whether somatic or germ line, not only to correct medically defined defects but also to enhance desirable traits—or even to introduce novel kinds of genes into people. Though the initial fear was that government might impose such a Brave New World, the greater danger would probably be the free market, with parents wishing to purchase—and hucksters willing to promise—beauty, athletic prowess, or intelligence for future offspring. However, for the foreseeable future there is a technical safeguard: the number of genes affecting each of these traits is large, and so it will be difficult to manipulate or even to identify them.

There are also social safeguards. With few historical exceptions, most societies have rejected positive eugenics based on the usual reproductive techniques, though it obviously could be just as effective in humans as in domesticated animals if the kind of selection were agreed on. It therefore seems unlikely that we would embrace use of elaborate molecular genetic techniques for the same purpose. On the other hand, we cannot dictate or predict mores for many generations ahead. For our own time we can hold the line and limit genetic engineering, both somatic and germ line, to the correction of disease. Yet growing understanding of polygenic traits may blur the line between preventing disease and enhancing well-being.

Meanwhile, problems are virtually guaranteed in the near future from another advance: greatly improved ability to diagnose genetic defects. Used prenatally, to provide the option of preventing the birth of individuals with serious defects, this practice has been the greatest contribution of clinical human genetics to decreasing misery. But the advances in molecular genetics that will increase its power will also improve our ability to detect defective genes after birth, and here the benefits are more ambiguous. On the positive side, detection of carriers can provide guidance for their future pregnancies, and in some diseases early diagnosis in the newborn may lead to preventive or ameliorative measures. On the other hand, screening various populations for specific genetic defects will create societal problems. These include painful predictions of deterioration of health without hope of therapy or prevention; invasion of privacy; and obstacles to employment and to insurance, both realistic and exaggerated by misinterpretation. However, the insurance problem may also have an indirect benefit: since the proliferation of predictive information will complicate private health insurance, it may lead to strong pressure—probably after a period of abuse—for universal health insurance.

It seems clear that we have no realistic choice but to continue to study genetic factors in disease. We must also try to avoid inappropriate responses, such as screening for the sickle cell gene—once widely legislated, but now abandoned because the information did more often harm than good. Friedmann concludes that we shall probably use our genetic knowledge both well and badly.

Chapter 10: Neuroscience

As a background for discussing the contributions of molecular genetics, James Schwartz presents a succinct, scholarly description of the basic features of the nervous system. He deals not only with its exceptionally specialized set of structures and functions, but also, briefly, with the challenge, shared by psychology and philosophy, of trying to understand consciousness and cognition.

The chapter starts by reviewing the prolonged resistance to the idea that different mental functions are localized in different parts of the brain. This principle was advanced early in the nineteenth century by Franz Gall, but it was unfortunately discredited by giving rise to a field, phrenology, that claimed to infer individual propensities from the shape of the cranium.

The next sections of the chapter describe the structure and activities of the basic unit of nerve function, the neuron. In this cell a main extension, the axon, forms an effective contact (synapse) with a distant target, while shorter branches form additional synapses with as many as a thousand other neurons, providing an intricate network of communication. A nerve impulse is conducted along a nerve cell as a wave of transiently decreased electrical potential across the membrane, caused by reversible, brief changes in the flow of various ions. When this "action potential" reaches a synapse it causes the release of stored neurotransmitter molecules, which may be either excitatory or inhibitory; the response of the target cell is determined by the balance of the two kinds, received from multiple neurons.

Like the switches in a computer, synapses provide the basis for memory (learning) by undergoing changes that modulate their future transmission. Molecular genetics has opened this field to experimental analysis, since the ability to clone the genes for synaptic proteins has made it possible to identify them and to demonstrate modifications associated with memory.

Neuroscience also studies the complex integrative activities of the brain, seeking to locate the sets of neurons ("nuclei" or centers) involved in various functions and to elucidate the connections among them. So far molecular genetics has had little effect on this area, compared with the neuron. However, identification of key genes and proteins promises striking advances, in the next few decades, in understanding how specific axons find their way in the embryo to the correct targets, within the brain or outside—the remarkable process of building up the intricate circuitry of the 10 billion neurons of the human nervous system.

Neuroscientists have already made some progress in the study of this developmental process. Proteins secreted by certain brain cells are required for growth of the brain in embryos (and they offer promise of promoting regeneration of connections in adults after brain damage). Specific proteins of the cell surface are responsible for adhesion between particular cells. Finally, molecular techniques can be applied even to the study of localization. Thus, certain peptide synaptic transmitters (short chains of amino acids) are specific for pain, and they are synthesized via a messenger RNA, which can be detected by hybridization with a radioactively labeled, complementary nucleotide sequence. This development provides a novel way to identify brain cells involved in recognizing pain impulses.

Schwartz offers a rather critical view of the Panglossian assumption that increased understanding is always a good thing. He worries that we may miss something if the success of molecular neurobiology discourages other, more traditional ways of trying to understand consciousness, a fundamental mystery of life. I agree with his intent to oppose an excessive reductionism. But I would emphasize the distinction between the content and the mechanisms of consciousness. We can hardly expect to specify in terms of detailed neural events the rich flow of information through our conscious minds, and its fluctuating directions: here reductionism ends and humanistic studies begin. On the other hand, the fundamental nature or mechanism of consciousness seems likely to become accessible as a problem in neuroscience, rather than remaining in philosophy.

Without pretending to any deep insight into this ancient problem, I feel obligated to amplify this presumptuous suggestion. I would suggest that consciousness has evolved as a feature of the ability of an organism to *choose* its response to a set of stimuli, whether external or endogenous, and thus to make a *decision* rather than responding automatically by reflex. In reaching a decision the organism must integrate three classes of input: its exogenous or endogenous stimuli; the pertinent facts and conflicting values that it selects, through a chain of associations, from its body of stored knowledge (genetic and learned); and its current emotional state. This process of integration and decision making must involve certain centers common to all conscious decision making, as well as the variety of additional, alternative centers that can be tapped. Though there is no homunculus within our brain directing our consciousness, there must be a specific circuitry, partly fixed and partly fluctuating.

The acts of integrating information and then making a decision are in effect a definition of consciousness—for one cannot choose or decide without awareness. We would thus have to attribute consciousness to any animal that is free to make a decision between alternative courses of action: a mouse, but probably not a fly. Understanding the physical basis of consciousness at a certain level does not seem beyond the reach of neuroscience. We have already identified the centers for sleep, which switch consciousness off and hence must be involved in switching it on. But dissecting the process that then guides the flow of thought will surely be much more difficult.

Chapter 11: Evolution

Francisco Ayala focuses mostly on our revolutionary new ability to use molecular sequences to reconstruct the complete evolutionary history of organisms, including the human species. The chapter begins, however, with the highlights of the earlier Darwinian revolution.

Until Darwin the scientific world suffered from conceptual schizophrenia: the world of nonliving matter could be explained by natural laws, but the origin of organs exquisitely adapted to the needs of living organisms required supernatural explanations. Darwin's theory completed the Copernican revolution by subjecting the origin of the living world to natural laws as well. It evoked religious opposition not primarily to the idea of evolution itself, which was not new, but to natural selection, a mechanism that excluded God from the process.

The evidence for evolutionary continuity of the whole living world came mostly from paleontology and comparative anatomy. Comparative biochemistry later provided further evidence: all organisms are remarkably similar in their mechanisms of energy production and their building blocks. However, biochemistry did not have a major impact on the advance of evolutionary biology until nucleic acid and protein sequences made it possible to quantify the evolutionary relations between species, including even the most distant ones.

Ayala uses cytochrome c, a component of energy metabolism, to illustrate this approach. This protein changes very slowly in evolution, and so it is especially useful for comparing distant organisms. To compare closely related organisms one uses proteins that evolve more rapidly (because their functions can survive a wider range of mutational changes).

In addition to refining the evolutionary trees, molecular studies have revealed an unpredicted feature of evolution: that gene duplication plays a key role in creating novelty, including formation of larger genomes as organisms evolve greater complexity. Thus, when a cell generates a second copy of an essential gene, one copy can retain the initial function in subsequent generations (because cells with mutations that destroy the function do not survive), while the other copy can accumulate mutations and thereby evolve a sequence with a new function. Accordingly, as computer-aided comparisons of sequences have shown, the proteins within an organism, as well as homologous proteins in different species, can be classified in a limited number of families, each descended from a common molecular ancestor. Indeed, almost any study of a new protein now includes this kind of evolutionary comparison—a molecular comparative anatomy.

Taking up another problem, Ayala notes that resistance to Darwinian evolution has arisen not only from religion but also from conflicts with common sense: it was hard to accept the idea that such a complex organ as the eye could have evolved in tiny steps (now identified with mutations) if that evolution depended on improved fitness at each stage. However, molecular genetics has now moved toward solving the problem by showing that there are many classes of mutation, in terms of both the kind of molecular change and the magnitude of the phenotypic effects.

The smallest molecular change, the replacement of a single nucleotide, can have a variety of phenotypic effects, some much larger than alteration of the activity of one enzyme. For example, an alteration in a regulatory sequence may affect the formation of many proteins dependent on that regulator. Or addition or removal of one nucleotide within a gene causes a "frameshift" in the reading of the subsequent sequence of nucleotides in units of three (codons), so that sequence now codes for a completely different set of amino acids.

In addition, evolution has much larger one-step changes in DNA to build on. Among these, large blocks of DNA can be transferred to new locations within the genome of an organism, or between the closely related plasmids. Another process accelerates the formation of useful novel genes by recombining pieces of earlier genes, using the same successful "motif" over and over.

Evolutionary biology, like genetics, has moved from formal relations to concrete, demonstrated mechanisms. But it still faces broad challenges, such as the need to explain how the 1 percent difference in sequence between humans and chimpanzees can result in their large phenotypic differences. Here molecular approaches to developmental biology will play a large role. Another challenge is to provide a reasonable explanation for the "origin of life" from inanimate matter—more properly designated as the origin of genetic information, since the process probably involved a gradual increase in complexity and information transfer accuracy, without a sharply defined moment of birth of the first cell.

Evolutionary biology thus plays a key role in understanding some practical problems, such as the possible spread of engineered organisms and the value of species conservation. It has thereby become more relevant to practical and political affairs. Ayala ends with a plea that the subject be given the attention in schools that it deserves, despite the objections of fundamentalists.

Human Genetic Diversity

I shall close by noting a more distant but potentially major impact of molecular genetics, related to neuroscience and to evolution but not discussed in those chapters: its effect on our understanding of genetic aspects of individual differences in cognitive and other behavioral traits. Current behavioral genetics, dealing with phenotypes but unable to identify specific genes, has been a highly controversial area. And unfortunately much of the debate has subordinated scientific fact to moral or ideological conviction. Since this process is the obverse of what philosophers have called the naturalistic fallacy, that is, trying to derive an "ought" from an "is," I have tried to identify it more sharply by calling it the "moralistic fallacy."[4]

It seems almost inevitable that in the next few decades neuroscience will begin to correlate individual mental differences with molecular differences in the brain. For example, differences in speed of mental operations might be due in part to differences in specific synaptic proteins that affect the speed of transmission of information (like differences in the switches in a computer), as well as differences in factors that regulate and direct a more complex set of processes, the formation of the various centers in the developing brain. Such objective information should help us to understand and to build on the realities of human diversity. And while reliable insights in this area are still distant, the extraordinary rate of current progress in molecular biology suggests that the time scale may surprise us.

Molecular genetics may also contribute indirectly to the recognition of genetic diversity by promoting acceptance of evolutionary principles. For all genetic diversity has arisen through the mechanisms of evolution—and while in humans the genetic differences are more readily demonstrable for physical and biochemical than for behavioral traits, the latter are subject to the same evolutionary rules, which predict broad diversity. Indeed, the evolution of our species has been especially rapid with respect to its behavioral capacities, tripling the size of the brain within a mere 3 million years; and this development obviously required the continuing formation of a large reservoir of genetic diversity on which natural selection could act. Moreover, like the diversity in resistance to infections, which has ensured that even in the worst plagues some individuals survived in virtually every community, the diversity in our various talents, drives, and propensities must also have had great survival value in our earlier evolution. It clearly has great cultural value today. What an inefficient and dull society we would have if the human species were a giant clone, instead of having a population with diverse talents, and a unique personality in each new baby!

But however objectively we learn to distinguish the genetic from the cultural components of our diversity, it will not be easy for us to accommodate this knowledge, because it will intersect intimately with our values and our concepts of justice. Here science cannot prescribe solutions. But it can be a critic of the assumptions on which we build policy. As was emphasized by Theodosius Dobzhansky—one of the great evolutionary biologists, and Ayala's teacher—human equality is a moral and political ideal, which aims at increasing equality in various social spheres that we can organize; it must be sharply distinguished from the notion of biological equality, which we cannot legislate. We distort the goal, and we jeopardize its realization, by trying to strengthen it with an artificial and vulnerable empirical base, built on the false assumption of biological uniformity.[5]

The penetration of molecular genetics into this area will surely generate fresh tensions. Robert Weinberg has spelled out the problem eloquently, in discussing the problems of the Human Genome Project:

> These discoveries have the potential of affecting reproductive decisions, education policy, hiring practices, insurance practices, marriageability, and may even lead, more ominously, to discovery of genetically templated differences between different ethnic groups and races. This means that the long-term effect of genetic mapping techniques will have a societal impact that will greatly overshadow the influences of all other aspects of the revolution in biology and the associated biotechnology. We have not yet begun to seriously ponder the demons issuing forth from the Pandora's box that we are about to open.[6]

When we do have to face these questions in the future some persons will no doubt see our increasingly detailed information about genetic differences as a wonderful basis for tailoring education better to individual needs, and for creating more effective ways to promote fulfillment of individual potentials. Others will focus on the tragic consequences of earlier premature applications of genetics to social policy by eugenicists, and the gross distortions of genetics by racists; and they will argue that the findings of molecular behavioral genetics, however objective, will inevitably lead to similar abuses. This controversy may prove much harder to resolve than our current arguments over the limits to genetic intervention in human beings, or over the introduction of genetically engineered microbes into the environment.

Notes

1.   B. D. Davis, "Evolutionary Principles and the Regulation of Engineered Bacteria," *Genome* 31 (1989): 864-69.

2.   News report, *ASM News* (Am. Soc. Microbiol.) 56, no. 8 (1990): 407.

3.   B. D. Davis, "Limits to Genetic Intervention in Humans: Somatic and Germline," in *Human Genetic Information: Science, Law, and Ethics*, Ciba Foundation Symposium No. 149 (Chichester: John Wiley, 1990) pp. 81–92.

4.   B. D. Davis, "The Moralistic Fallacy," *Nature* 272 (1978): 390.

5.   T. Dobzhansky, *Genetic Diversity and Human Equality* (New York: Basic Books, 1973).

6.   R. A. Weinberg, "There Are Two Large Questions," *FASEB J.* 5 (1991): 76.

15

Comments: Public Policy Issues

Harvey Brooks and Rollin B. Johnson

This chapter will comment on the discussions, in various chapters, that bear on policymaking. It will especially consider Aaron Wildavsky's view, expressed in Chapter 6, that differences in the assessment of risks and benefits persist not simply because of individual differences in ability to grasp and assess the evidence: they reflect fundamentally conflicting views about our social structure, relationship to nature, responses to uncertainty, and visions of the future of our species, our culture, and the planet.

Underlying Assumptions

Wildavsky argues that for reasons of political expediency, both the proponents and the opponents of biotechnology take more moderate stances on regulatory policy than the logical implications of their beliefs would demand.[1] These implications, he argues, should have led the proponents to work for the abolition of regulation, and opponents to insist on the abolition of this research and its applications. Of course, this dichotomy is artificial and extreme; but it does draw attention to real features of the controversy that lie beneath the surface.

Wildavsky thus focuses on those critics who emphasize the "monstrous potential of biotechnology—in disasters so bad that no good . . .

could possibly overcome them." This anxiety was initially widespread. It may still motivate a small minority of critics, and an unknown fraction of the public—aware of the plague of rabbits in Australia initiated by a single pair, and fearful of even greater catastrophes analogous to Chernobyl or Bhopal. And science cannot prove the impossibility of this kind of rare event. However, at the end Wildavsky provides many reasons for believing that in practice any attempt at absolute prohibition of biotechnology could not work. It would merely drive the technology underground or into less developed countries, where more limited resources for assessment and regulation would even increase the net societal risk.

In fact, most critics are concerned with a rather different, less extreme kind of risk: not catastrophe, but a "slippery slope" in the interactions between the socioeconomic system and evolving technological opportunities. A series of steps may give rise to undesirable technological, social, and economic changes, though these steps would not have been allowed had the implications of the technology been fully debated and understood at the outset.[2] In the words of Margaret Mellon (Chapter 5), for whom this is a primary concern, "Each of these developments represents a choice, not an inevitability," and "environmentalists have joined the [biotechnology] debate . . . to bring an environmental perspective to these choices." She cites the many "miscalculations of the past . . . constantly before the environmental community," in which "miracle" technologies offering great benefits to humanity later revealed their dark side. The effects include "polluted air and water, pesticide-contaminated ground water, manufacturing and transportation accidents, ten thousand abandoned hazardous waste sites, an eroding ozone layer, and the possibility of rapid global climate change."

A major reason for these failures is timing. The benefits of a new technology appear immediately, and they increase as its use expands. But the deleterious side effects (especially of "polluting technologies") appear late—either after a long latency period (as with carcinogenic substances) or when the scale of use exceeds a threshold (as with the smog arising from automobile exhausts). Thus, the negative impacts of a technology often become apparent only after it has been widely deployed and society has become highly dependent on its benefits. By then the cost of abandoning the technology may have become prohibitive.

This last point argues for a kind of careful, anticipatory regulation of the new technology, and a conservative attitude toward its acceptance. This argument cannot be lightly dismissed, but it misses something important. It does not acknowledge the severe limits of our ability to anticipate the consequences of a new technology. Moreover, it does not attach sufficient weight to the opposite risk: that overregulation and bureaucratic procedure can forfeit or delay important benefits. For these reasons Wildavsky suggests an alternative approach, which he calls *resilience:* careful monitoring for signs of danger, and preparation for rapid response when they do appear, rather than the pretense that we know enough to be able to prevent them altogether.

We would agree that excessively conservative regulation is not justifiable. But neither is "full speed ahead," with no regulation (if any support that view). A more logical basis for regulation is put forward by Henry Miller and others[3] in an article that expands upon his chapter (Chapter 12) in this volume. These authors put forward a tentative systematic decision "algorithm" or "coordinated framework" for determining eligibility for either regulation or exemption. The current debate would surely be much more profitable if directed at the specifics of such a framework and its technical logic, rather than at general considerations of the "strictness" or "looseness" of the regulatory regime.

In this connection we must recognize that the impact of regulations on technical progress does not depend simply on how strict or relaxed they are. It also depends on the predictability of regulatory decisions; and these are obviously more predictable when they can be based on rational and well-understood scientific principles. Moreover, regulations cannot deal with biotechnology as a unit: as Miller's chapter emphasizes, the processes of evaluation are quite different for different classes of products (e.g., pesticides, hormones) and for different kinds within a class.

The fears of the critics are further exacerbated by what appears to be an exaggerated view of the "technological imperative," a belief that if something can be done it will be done. In actual fact scientific research turns up far more theoretical possibilities for applications than can possibly be pursued, no matter how great the resources available. Hence we cannot escape the need to make discriminating technological choices. Attempting to prevent possible misapplications of new knowledge by not seeking the knowledge itself in the first place throws out the baby with the bath, for it forecloses too many opportunities for benefit.

In the context of these arguments there appears to be ample room in principle for compromise and consensus between all but the most extreme critics and proponents of biotechnology. In practice, however, consensus for specific actions may be hard to implement. In addition to hidden agendas having to do with the relative power and financial support of various disciplines and organizations, problems arise even more from differences in implicit "world views."

Rival Cultural Constructions

In many ways the most interesting and challenging contribution of Wildavsky's chapter is his typology of "world views," which he considers to be more or less immutable, and his contention that they predict an individual's positions on biotechnology regulation much more than do differences in awareness or understanding of specific evidence or arguments. He is vague, however, as to how these world views are formed and why they are apparently so immutable. Are they based on total life experience? Are they deeply ingrained in the personality as a consequence of inheritance or of early socialization, and thereby passed on from parent to child?

This model was initially based on other areas of risk management such as nuclear power, but it has a great deal of explanatory value for the genetic engineering debate. It is supported by the finding that in areas of scientific uncertainty, as they relate to policy options, there is frequent and striking clustering of different expert views on specific issues. For example, in the debate over whether to continue with the construction of the Clinch River Breeder Reactor, during the Carter administration, the experts (regardless of discipline) seemed to divide into two groups.[4] The proponents thought that electricity consumption would grow rapidly in the next twenty-five years; that the supply of usable uranium ores was relatively low; that future costs of photovoltaic cells to recover solar energy would be relatively high; and that we could prevent diversion of plutonium from nuclear fuel processing into nuclear weapons proliferation. The opponents offered estimates clustered at the opposite end of the plausible range of values. Nobody offered a hybrid view, expecting photovoltaics to be cheap and uranium supplies ample, even though these two judgments were not technically linked and hence should not have been correlated in the judgments of objective experts. There were evidently systematic biases at work based not on the real technical issues but on policy preferences.

This example fits, though it does not prove, the typology of social constructions of reality proposed by Wildavsky. Briefly, he divides these into four visions of linkage between the natural and the social world. The first links the belief in a "cornucopian" nature with "individualists," who believe that competitive enterprise and ingenuity can always surmount constraints imposed by nature. Next, "hierarchists" see a rather similar "tolerant" nature and emphasize the role of experts in overcoming its constraints. "Egalitarians" see a "fragile" nature: hence collective social constraints are needed to prevent those in power not only from oppressing the weak, but also from destroying the delicate balance of nature and thus triggering its vengeance. Finally, "fatalists" see a "capricious" nature on which we have little influence, because it is essentially random and unpredictable; we can only roll with the punches.

This provocative characterization may help us to understand the controversies over genetic engineering, even though it is undoubtedly a gross overgeneralization, reasonably summarizing some observations about the real world but not others. Thus, while it is easy to recognize various aspects of nature as cornucopian, fragile, or tolerant, nature taken as a whole can hardly be described in this way. Many so-called nonrenewable (i.e., fragile) resources may actually be cornucopian because of technological progress that substitutes more abundant raw materials for scarcer ones—for example, iron for bronze, or successive new energy sources. On the other hand, certain theoretically renewable resources, such as biological species, may suffer extinction or otherwise exemplify fragile nature. Furthermore, nature is not monolithic: some ecosystems, either natural or constructed, are robust; others are fragile. However, Wildavsky might reply that even so, individuals may hold a monolithic view of nature.

The actual organization of any national society probably represents a more or less precarious equilibrium among all the world views described by Wildavsky. However, this theory gives very little guidance to understanding how this equilibrium might evolve. What factors determine the emergence of a dominant world view in any particular epoch? In what ways, if any, does the popularization of contemporary science influence the struggle between contending world views? And how stable or malleable are the world views held by individuals, or by groups?

In principle, generalizations about both the natural and social world should be amenable to empirical test, by scientific observation and by historical description or analysis. But Wildavsky denies that people generally engage in such testing: they inextricably mix policy preferences with alleged descriptions of the real world, one supporting the other. And even though scientists have been socialized to deep respect for the goal of objectivity in their work, they are not necessarily less liable to bias in other areas. Certainly, one is struck by how the strong proponents of any view tend to talk only to each other.

Wildavsky's predictions, in discussing the future politics of biotechnology regulation, seem to be predicated on the universality of this kind of rigidity, resulting in resistance to reality testing or to contrary viewpoints. On the other hand, the rigid proponents of a view may be only a small, dedicated minority, while the majority may be more susceptible to persuasion or to empirical evidence. In a more extensive paper dealing with the same typology, Wildavsky concedes that "change occurs because the ways of life, though viable, are not impervious to the real world."[5]

If Wildavsky's typology is to help us understand the probable future evolution of the politics of biotechnology regulation, it seems to be crucial to understand how transitions among different world views occur. For example, are they mostly gradual, or do they occur suddenly in response to dramatic events? If they are gradual, then their susceptibility to influence by the scientific community is probably much greater. The best strategy for the scientific community might then be to emphasize that various world views are in fact overgeneralizations derived from exposure to limited segments of reality, and so we should try to separate concrete issues from very general philosophical positions. Indeed, perhaps all that scientists can do, as scientists, is to offer evidence on concrete issues, and to recognize that groups with different backgrounds will respond with different thresholds or at different rates.

In an encouraging development, a recent study by the Public Agenda Foundation demonstrated that even on issues involving highly technical considerations, if a representative group of ordinary citizens takes the time to become informed their initially apprehensive views can be replaced by judgments not too different from those of a consensus of experts.[6] This experiment suggests that the average citizen has not tried to formulate an internally consistent and defensible world view and hence is open to change based on new information that he or she trusts. Wildavsky's typology may thus fit only a small set of individuals. However, these individuals have often played a major role as activists, stimulating concern and support from the rest of the citizenry.

One can perhaps relate Wildavsky's typology to Herbert Simon's concept of "bounded rationality."[7] In his view the human mind is not ordinarily capable of assimilating all the information relevant to complex issues, but instead makes heuristic simplifications based on past experience, rules of thumb, analogies, and simplified models of reality. Wildavsky's typology may be looked upon as one attempt to classify the heuristic simplifications found in practice.

Role of the Courts

This simplification of science in decision making clearly applies to the way science and challenges to science policy are treated in the courtroom. As noted in Richard Stewart's chapter, "law generally works best when differences of interest or value can be reduced to well-defined factual issues that are capable of being resolved through the adversary process." But when decisive evidence is lacking, "the law may be ill adapted to assess and manage biotechnology."

Facing this problem, legal processes search for a pragmatic compromise solution, while science attempts to settle disputes by seeking more decisive evidence. The law is time-bound; whereas scientists are profoundly interested in finding out where the truth about a technical issue really lies, the law cannot afford to wait for the truth to emerge over time. The suspension of judgment that is a normal part of the scientific process strikes a lawyer as being an expensive luxury.

Stewart underscores the importance of other cultural differences between the scientific and legal communities, and he discusses the tension between the fundamental principles of science and of law. The scientific community is accustomed to fashioning a consensus interpretation from diverse collections of data and competing models of nature. The focus is thus not on defending the efforts or opinions of individual scientists; it is on developing models that accurately mirror nature and so help us to predict realistically the consequences of alternative decisions. The courts, on the other hand, are constitutionally mandated to uphold the rights of the individual. And while broad judicial principles exist, they are often partially waived to fit the particular circumstances of each case.

This fundamental difference is most evident in the use of scientific evidence in the theater of the courtroom. For example, in response to DNA fingerprinting data for identifying criminals one federal court judge has recently stated that so powerful a technique "places a fist on the scale of justice" against the defendant, and so it should be held to higher standards of scientific agreement than more traditional techniques.[8] Since it is easy to garner conflicting testimony from scientific "experts" on almost any subject, and since there is almost never 100 percent certainty in the science that enters into judicial decisions, this requirement can hamstring science in the courtroom and confuse both judge and jury. As Stewart notes, in public suits against regulatory agencies this tension is normally worked out, not very satisfactorily, by a retreat into judgments of procedure, "requiring administrators to write down their decisional criteria and to follow them in agency opinions explaining why the experts' judgment was an appropriate basis for an administrative decision made in the face of adverse criticism."

Thus, the courtroom is not a good place to resolve technical issues. How, then, does the public have input into the policymaking process? Stewart believes that in the development of biotechnology the public's major impact may be its pressure on regulators to be conservative, in order to avoid future litigation. In areas where scientific certainty is perceived to be lacking, as in the introduction of genetically engineered organisms into the environment, public perceptions will play a large role, as will ethical judgments about where the burden of proof should lie. It is under these conditions that the public is most susceptible to demagoguery and to ideological arguments. And given the eagerness of regulatory agencies to assuage public concerns, it would seem to us desirable for these agencies to take responsibility for helping the public to become better informed.

Stewart further insists that in the face of uncertainty we must rely on informed scientific judgment about the probability and extent of harm, rather than demand unattainably rigorous inference. To this end greater input and visibility should be given to independent scientific boards advising regulatory agencies, in order to promote public confidence in the regulatory decisions that are made. This is not to say that the public can avoid responsibility, for it is always the ultimate judge of whether a technology should be accepted. Rather, what Stewart proposes is a means by which scientific evidence receives a proper hearing and the public is assured that decision making is carried out responsibly.

Ecologists versus Microbiologists

An interesting issue in this volume is the evolving dialogue between ecologists and microbiologists. Simon Levin's chapter represents a conscientious effort to bridge the gap between their perspectives. He emphasizes points of agreement and suggests directions in which to seek further convergence.

It is easy to understand how the different professional experiences of these two groups tend to reinforce their adherence to different overgeneralizations embodied in Wildavsky's world views. Ecologists readily see nature as fragile, because ecological systems are often interconnected in such a way that small disturbances can be magnified into big effects. But this is by no means invariably true; ecosystems, including those modified by humans, are often robust and resilient. Nevertheless, the fragile systems tend to be the most interesting to study, because they present the most challenging conceptual problems; rare experiences with the introduction of exotics thus attract an inordinate amount of attention and research.

Such cases emphasize the dependence of ecological processes on particularities, thus making ecologists very wary of generic statements about the properties of ecosystems. Levin further observes that ecologists often have to provide scientific advice not only in the face of uncertainty but in situations in which the prospective costs and benefits are not distributed uniformly across all segments of society. Levin thus fulfills Wildavsky's prediction that the fragile-nature world view would be linked with an egalitarian view of distributive justice.

Microbiologists are accustomed to dealing with a very different reality. Many of the phenomena that they study are common to all living organisms, and so the behavior of particular systems can often be extrapolated to other, far different systems. Furthermore, the ability of microbiologists to manipulate the microbial world in controlled ways reinforces the notion of mastery of a cornucopian or tolerant world, waiting to be exploited by bold individualists with expertise. In addition, the rapid succession of exciting discoveries by contemporary microbiologists and molecular biologists promotes an atmosphere of almost Darwinian competition in their research, whereas the ecologist is used to less spectacular progress in the study of systems with many uncontrollable variables. This experience fosters more humility about the unpredictable consequences of manipulating nature.

The basic point is that the professional experience and research lifestyle of different scientific disciplines, each focusing on limited segments of reality, strongly influence the world views of their members. Of course it is possible, instead, that the different world views are innate and determine an individual's choice of vocation. Most likely there is some two-way causality, each process reinforcing the other.

One could also speculate that in organizing the first Conference on Recombinant DNA Molecules, at Asilomar in 1975, the molecular biologists, influenced by the atmosphere of their campuses in the preceding decade, initially adopted some of the values implicit in the egalitarian world view. As Allan Campbell's chapter notes, they felt virtuous in sounding the alarm over their uncertainty. But it seems that the combination of their view of the natural world and their professional experience, reinforced by new evidence, soon led them to retreat from the initial alarmist position—though the public did not accept the reversal nearly as readily as it had the original alarm.

Though scientists are deeply dedicated to objectivity, the debate among the microbiologists and ecologists suggests that in generalizing from and interpreting the evidence, compared with collecting it, objectivity may sometimes be elusive. Nevertheless, in contrast to most earlier, rather descriptive work in biology, molecular studies can now often come up with strong, reliable inferences, much like physics, and these are relevant to predicting the properties of organisms with defined genetic changes—especially properties related to the crucial question of pathogenicity versus harmlessness. It is much harder to collect such clear evidence in ecology. Microbiologists therefore have greater confidence in predictions than ecologists do.

The pessimistic scenario is that ecologists will not be persuaded and microbiologists will not have patience with that position; hence the public and regulators will continue to be torn. In a more optimistic scenario—assuming that no significant damage appears—the problem will gradually evaporate on the basis of familiarity with the new technology rather than of any real accommodation. Our problem is how cautiously or boldly to proceed in the interim.

Questions of Scale and Pace

Apart from questions of risk, a separate set of arguments for caution invokes society's limited capacity to adapt rapidly to changes. The social and environmental risks that Mellon labels "broader concerns" are a function not only of the characteristics of the technology but also of its social and economic context. For example, if advances in biomedical technology are to realize their potential they often require changes in health care delivery—and as we have seen, changes in such infrastructures, deeply embedded in culture and history, tend to lag behind technological innovations.

The responsiveness of societal supporting systems is challenged further when the pace of technological advance increases dramatically. We have been able to live comfortably with the gradual pace of innovation seen in traditional selective breeding in agriculture; but the striking acceleration of the process by genetic engineering could create social problems, even though it has fundamentally the same aims. Thus, if several products of a new class are introduced at invervals there is opportunity to monitor for delayed side effects that could influence the development of the later members of the class; if they are all introduced at about the same time this opportunity is lost. Levin's chapter suggests an additional, hypothetical effect of rapid innovation: unexpected synergistic effects among several products. However, that seems unlikely for multiple novel organisms with distinct ecological niches and effects, in contrast to multiple chemicals that may affect a common process in organisms.

Throughout the chapters by Levin and Mellon questions of the effect of rate and of scale loom large. While both authors accept the consensus that regulation should be based on the product and its proposed use rather than on the role of genetic engineering in its production, the acceptances are somewhat cautious and grudging. The main reason is that genetic engineering permits the development of new products at a much higher rate than classical selective breeding, and so there is less time for observation and assessment of possibly unforeseen ecological effects. But even so, in the absence of positive evidence of risk it is difficult to see how this argument can affect our decisions, except to reinforce a policy of vigilant monitoring coupled to a capability for resilient response to new evidence.

A further concern is that a larger scale of introduction will increase the chance that genes for desired properties for particular species in local environments, such as frost tolerance or herbicide resistance, could spread to other species or over broader geographical ranges through natural mechanisms such as plasmid exchange. Still another fear is that economically or socially attractive applications of genetically engineered organisms may expand out of control, with ecological damage dependent on scale. A number of initially valuable chemicals provide models, such as DDT, polychlorinated biphenyls (PCBs), and chlorofluorocarbons (CFCs). Again, in the absence of positive evidence of toxicity, pathogenicity, or other harm from engineered organisms, in contrast to chemicals, it is difficult to see how to control or anticipate undesirable effects.

Mellon adds a new dimension to these concerns by focusing on the socioeconomic consequences of otherwise benign introductions: use of bovine growth hormone might displace traditional sources of livelihood, and use of crops or animals tolerant to various chemicals might encourage social tolerance for chemical pollution. The argument is that these innovations should be discouraged because society is not politically or morally "ready" for them. This position raises the intriguing issue of whether application of a new technology, admittedly beneficial in principle, should be deferred until society is "ready" for it, or whether its availability should be used to force a more rational order of society so that the benefits can be realized—for example, responding to bovine growth hormone by reevaluating milk subsidies.

This issue was explicitly faced in a confrontation in India over the proposal to introduce the new seed varieties of the green revolution during the famine of 1966.[9] The social scientists opposed it because they feared it would exacerbate income inequalities between large and small farmers. Agricultural scientists favored it. After listening to the debate Indian officials decided to allow the introduction. The social consequences were mixed: the changes did exacerbate income inequalities in some regions, but they prevented widespread starvation. Moreover, though the social scientists predicted that peasant farmers were too conservative to adopt the new agricultural technology, the prediction proved to be spectacularly wrong in most cases.

Neither Levin nor Mellon discusses the concrete policy implications of these broader social concerns, other than to give a general admonition in favor of regulatory conservatism. And in practice such conservatism today means continuing to require approval of test introduction on a case-by-case basis. The prospective development of thousands of new strains of organisms will then surely overwhelm the current regulatory system. As Stewart's chapter points out, three outcomes are possible: (1) continuation of case-by-case approvals; (2) cursory approvals; and (3) categorical exemptions of presumably harmless classes of organisms, based on increased knowledge and experience. The case-by-case approach would cause severe product delays, resulting in reduced investment, lower competitiveness, and the postponement of benefits to the public. Cursory approval, or rubber-stamping, would expose the public to "unscreened" products and hence perhaps a higher level of risk. Exemption of categories would allow rapid approval of large sets of experiments or products, providing time and resources for closer scrutiny of those with higher risk.

The last of these scenarios, according to Stewart, is eventually the most likely. We agree, but with no idea of the time scale. Meanwhile, those who defend case-by-case regulation recognize that current resources are not sufficient, and they recommend expanding these resources. Many observers also recommend devising legislation to create a single regulatory agency with a fresh mandate, replacing the use of existing legislation that had been created for other purposes. We suggest that it would be important to scale up resources—not so much to screen applications as to provide for increased monitoring and assessment, as experience expands, in order to recognize incipient dangers. As to the creation of a new unified agency, the apparent advantages of such a neat arrangement may be outweighed in practice by new problems, including disproportionate congressional accommodation to pressure groups. The idea of a single agency for biotechnology is very controversial. Large companies that are able to cope with existing mechanisms tend to oppose the risk attendant on major changes from the status quo, while smaller companies, without such legal resources, are more inclined to accept that risk in order to simplify the regulatory process.

The Burden of Proof: Level of Risk

In the early controversy over recombinant DNA critics demanded proof that the proposed experiments could not be dangerous. Such a search for absolute security is of course unrealistic, and the argument eventually became one of defining an acceptable level of risk and devising appropriate regulatory mechanisms. When the problems were later extended from the laboratory to field testing, microbiologists clashed not only with concerned members of the public but also with ecologists.

Levin emphasizes the increasing convergence between these two groups of scientists with respect to biotechnology regulation. But as Bernard Davis notes in Chapter 14, there appears to be a conflict here between appearance and reality. Not only does Levin continue to defend case-by-case regulation, but where we do find apparently increasing agreement it seems to depend on the use, by both sides, of qualitative phrases like "negligible risk," "highly unlikely" events, or "minimizing unnecessary regulatory burdens." These reflect the context of scientific uncertainty. And even Levin, not to mention the members of the Ecological Society of America panel whom he quotes, might differ from various members of the National Academy of Sciences panel about what probability constitutes a "negligible risk" or a "highly unlikely" event, or what *quantitative* level of risk makes a regulatory burden become "unnecessary."

We are therefore dealing with the subjective probability associated with such terms. Anything that has not been done before always presents some potential risk, if only because of scientific uncertainty or incomplete information. But one panel might consider a chance of error of one in a thousand quite acceptable, while the other might demand a chance of one in ten thousand or even one in a million. Yet any one of these numbers might be compatible with the qualitative terminology used in the reports issued by each panel. Peace and harmony thus still seem elusive. However, the work of Miller and his colleagues[10] in developing a systematic framework for the triggering of regulatory consideration indicates one direction in which progress can be made.

Costs of Delay

One theme that comes through strongly in several of the chapters is the high social costs of delay in research with genetically engineered organisms. In delaying we forgo not only novel products with social benefits, but also improvements in existing processes (in manufacture or agriculture) that would ultimately benefit the consumer through lower prices. Another cost is prolongation of current toxic chemical inputs into the environment.

Delay may also exact costs in the ability of new biotechnology companies to attract investors. Small companies may not be able to survive regulatory delays, while large ones may shy away from a product for which approval is uncertain or public controversy is likely. The morale and motivation of scientists seeking useful new technologies are also bound to suffer. And the effects may extend beyond technology, for much of the applied research in molecular and cell biology in industry contributes to the knowledge base on which all the life sciences and medicine draw. This cumulative knowledge forms a fabric of understanding that far exceeds the sum of the individual threads of information. Knowledge from carefully monitored field tests might be more valuable than laboratory experiments of ecologists, especially if the data and insights are open to continuous expert scrutiny (and public debate).

Unfortunately, the philosophy of encouraging experimentation (though with caution) has not always animated the debates in other areas of environmental regulation—a point not sufficiently stressed in Stewart's review of environmental regulation. In particular, though the requirements for an environmental impact statement (EIS) under the National Environmental Policy Act might be expected to focus on evaluation of a specific proposed experiment or prototype, judicial interpretation has tended to insist that the EIS go further and evaluate the impacts of potential subsequent systems. For example, the Clinch River Breeder experimental reactor was a single reduced-scale prototype, but its EIS was required to analyze the consequences of the fully developed international "plutonium economy" that might arise if breeder technology were deployed on an international scale. These distant issues included weapons proliferation, the large-scale transport and disposal of high-level radioactive waste, and security.

So far this legal theory has not been explicitly applied to the field testing of genetically engineered organisms. Yet the engineered Ice⁻ bacterium was opposed more as an entering wedge than as a menace— and the EIS required by the courts cost the University of California about half a million dollars, though there was no reasonable basis for doubt that the organism, already abundant in nature, was harmless. Proliferation of such impediments would be a further step toward the double-bind situation described by both Wildavsky and Stewart, in which insufficient knowledge prevents approval while fear of risk blocks efforts to obtain the required knowledge.

The legal theory would also erode resilience in response when a real problem arises—though the latter approach protects better against unforeseen risk, in our view, than reliance on the screening of individual projects on the basis of tests under a limited variety of conditions. Moreover, as Levin and Campbell both point out, if we accept the principle of basing regulation on product and not process, the debate over genetic engineering may lead logically to improved regulation of large-scale uses of the products of traditional techniques of genetic manipulation of bacteria. However, there is also danger of overregulation of classes of organisms that have been used for decades without harm.

Of course, one could argue that it does not make *economic* sense for government or industry to support an expensive experiment, prototype test, or field trial if environmental considerations (real or hypothetical) are likely to preclude ultimate operational deployment. In biotechnology this issue has not attracted much attention, perhaps because field trials are small and inexpensive (compared with the prototypes in some physical technologies), and perhaps also because it is hard to project the impact of large-scale use. Nevertheless, it is clear that uncertainty about the future is adversely affecting the eagerness of industry to enter agricultural biotechnology.

Conclusion

The difficulty in reaching consensus on issues in genetic engineering may reveal fundamental differences in views of the world that are hard to resolve—perhaps because they are buried deep in people's consciousness. These differences are seen in scientists as in anyone else, especially with issues that are even more emotionally charged than engineered organisms in the environment—such as the influence of genetic factors on intelligence or on criminal tendencies.

Fortunately, scientists are socialized with a strong ethic of building on objective evidence. As with other people, this does not always override considerations of personal ambition, political preference, or overall mindset. But scientists do gather and present empirical data, which eventually provide the firmest basis we have for judgment, even when they do not provide certainty. If education can promote a wider public search for informed judgments of the alternatives and their risks and benefits, then regulators, scientists, and the public should be able to arrive at a consensus on appropriate policies and actions. Whatever the rate of these accommodations, nature will inevitably determine the truth—although it may take a long time before it prevails.

Two strategies may assist in this process. One is to try to separate concrete decisions from overarching philosophical views of the world and of technology, and hence to focus on the possible implications of specific decisions in specific circumstances. The other is to attempt to arrive at an understanding of the reasons, in terms of both evidence and values, for differences among scientists in their interpretations of existing evidence and in the policy inferences supposed to derive from each. In this connection it would be especially useful to try to agree in advance on what new evidence or analysis would convince either side to change its mind.

Experience with successful public dialogues among scientists and laypersons, even after activists have stirred up anxiety, gives some reason for optimism about the possibility of arriving at consensus on such controversies. However, the biggest obstacle to consensus—both between scientists and the public and between scientists from different disciplines—is likely to continue to be the issue of where the burden of proof should lie in the face of inevitable scientific uncertainty. Should scientists be asked to prove the impossibility of varius hypothetical scenarios? Or should those who express concern be obliged to demonstrate real though low risks? Or should the two sides, while disagreeing as to whether the risks are negligible or are very low, nevertheless be willing to settle for better monitoring rather than for the much larger costs of stricter regulation? Without that solution the public will continue to be troubled by lack of confidence and excessive anxiety, and regulatory agencies will be troubled by these unsettling public perceptions as well as by the scientific dissension.

Notes

1. The draft title of Wildavsky's chapter was more revealing of the theme: "Goldilocks Is Wrong: In Regulation of Biotechnology Only the Extremes Can Be Correct."

2. R. A. Bauer, S. Rosenbloom, and L. Sharpe, *Second-Order Consequences: A Methodological Essay on the Impact of Technology* (Cambridge, Mass.: MIT Press, 1969).

3. H. I. Miller et al., "Risk-Based Oversight of Experiments in the Environment," expanded version of paper in *Science* 250 (1990): 490–91.

4. A. S. Manne and R. Richels, "Probability Assessment and Decisions Analysis on Alternative Nuclear Fuel Cycles," in *National Energy Issues: How Do We Decide?* ed. R. G. Sachs (Cambridge, Mass.: Ballinger, 1980).

5. A. Wildavsky and K. Drake, "Theories of Risk Perception: Who Fears What and Why?" *Daedalus*, Fall 1990, pp. 41–60.

6. Public Agenda Foundation, *Science and the Public Study*, vol. 1, "Searching for Common Ground on Issues Related to Science and Technology" (New York: Public Agenda Foundation, 1990).

7. H. Simon, *Models of My Life: An Autobiography* (New York: Basic Books, 1991).

8. "Justice System Takes a Second Look at Scientific Evidence," *New York Times*, 24 February 1991, p. E6.

9. W. D. Hopper, "Distortions of Agricultural Development Resulting from Government Prohibitions," in *Distortions of Agricultural Incentives*, ed. T. Schultz (Bloomington: Indiana University Press, 1978).

10. Miller et al., "Risk-Based Oversight."

Additional Reading

Books

Jackson, D. A., and S. P. Stich, eds. *The Recombinant DNA Debate.* Englewood Cliffs, N.J.: Prentice-Hall, 1979. An early, responsible presentation of a variety of points of view on the original intense controversy over experimentation with genetically engineered organisms.

Judson, H. F. *The Eighth Day of Creation.* New York: Simon and Schuster, 1979. A superb account of the development of molecular biology before recombinant DNA, emphasizing the discoveries and the personalities of key contributors.

National Academy of Sciences. *Introduction of Recombinant DNA–Engineered Organisms into the Environment: Key Issues.* Washington, D.C.: National Academy Press, 1987. An important, short report that defined and discussed the issues in a balanced way but did not succeed in defusing them.

National Research Council. *Field Testing Genetically Modified Organisms: Framework for Decisions.* Washington, D.C.: National Academy Press, 1989. A more detailed analysis, similar in thrust to the preceding NAS report.

Nichols, E. K. *Human Gene Therapy.* Cambridge, Mass.: Institute of Medicine, National Academy of Sciences; Harvard University Press, 1988. A summary, for a lay audience, of the proceedings of a symposium.

Nossal, G. J. V. *Reshaping Life: Key Issues in Genetic Engineering.* London: Cambridge University Press, 1985. A short paperback, with the excellent insights and judgments of an eminent Australian immunologist.

Office of Technology Assessment, U.S. Congress. *Field-Testing Engineered Organisms: Genetic and Ecological Issues.* Washington, D.C.: U.S. Government Printing Office, 1988.

Office of Technology Assessment, U.S. Congress. *Mapping Our Genes— The Genome Projects: How Big, How Fast?* Washington, D.C.: U.S. Government Printing Office, 1988.

Perpich, J. G., ed. *Biotechnology in Society: Private Initiatives and Public Oversight.* New York: Pergamon, 1986. A set of sophisticated essays on the social problems of biotechnology.

President's Commission for the Study of Ethical Problems in Medicine and Biomedical and Behavioral Research. *Splicing Life.* Washington, D.C.: U.S. Government Printing Office, 1982. A document that did much to define the issues in gene therapy and to promote a realistic discussion.

Science, Law and Ethics. Ciba Foundation Symposium 149: Human Genetic Information. Chichester: Wiley, 1990. A collection of thoughtful chapters on future prospects and problems, mostly by Europeans.

Suzuki, D., and P. Knudtson. *Genethics: The Clash Between the New Genetics and Human Values.* Cambridge, Mass.: Harvard University Press, 1990. A very readable popular account of the science and the social issues, presented from a left-wing perspective.

Zilinskas, R. A., and B. K. Zimmerman, eds. *The Gene-splicing Wars: Reflections on the Recombinant DNA Controversy.* New York: Macmillan, 1986. An additional series of essays with a variety of points of view.

Articles

Brill, W. J. "Why Engineered Organisms Are Safe." *Issues in Science and Technology* (Spring 1988): 44–50.

Cohen, C. "Restriction of Research with Recombinant DNA: The Dangers of Inquiry and the Burden of Proof." *Southern California Law Review* 51 (1978): 1081–1113.

Davis, B. D. "Bacterial Domestication: Underlying Assumptions." *Science* 235 (1987): 1329–1335.

Davis, B. D. "The Human Genome and Other Initiatives." *Science* 249 (1990): 342–343.

Watson, J. D. "The Human Genome Project: Past, Present, and Future." *Science* 248 (1990): 44–55.

Index

Numbers in italic designate illustrations or tables.

LIBRARY OF CONGRESS CATALOGING-IN-PUBLICATION DATA

The Genetic revolution : scientific prospects and public perceptions / edited by
  Bernard D. Davis.
      p.   cm.
  "Sponsored by the American Academy of Arts and Sciences."
  Includes bibliographical references and index.
  ISBN 0-8018-4235-2 — ISBN 0-8018-4239-5 (pbk.)
    1. Genetic engineering. 2. Biotechnology. I. Davis, Bernard D., 1916–
II. American Academy of Arts and Sciences.
QH442.G463   1991
660'.65—dc20   91-11019